Rumpler Aircraft of WWI

A Centennial Perspective on Great War Airplanes

Jack Herris

Great War Aviation Centennial Series #11

This book is dedicated with gratitude to Rumpler expert Reinhard Zankl, who provided key information and photographs for this book as well as other books in this series.

Acknowledgements
My sincere thanks to Colin Owers, Bill Toohey, Aaron Weaver, Reinhard Zankl, Lance Bronnenkant, Dave Hooper, Richard Andrews, and Milosz Zielinski for photographs, and Greg VanWyngarden for photographs and helpful suggestions and material. Thanks again to Reinhard Zankl for his help with serial numbers and to Colin Owers for his information on Rumpler postwar use. Thanks to Bob Pearson for his color profiles, Aaron Weaver for his digital photo editing and cover design, Martin Digmayer for scale drawings, and the Museum of Flight in Seattle for photographs. These books could not be produced without the support of the generous people acknowledged above who provided so much help and so many photographs. Any errors are my responsibility.

Thanks to Steve Anderson for the cover painting. Please see his website: **www.anderson-art.com**

I am looking for photographs of the less well-known German aircraft of WWI to illustrate future titles. To help with photographs please contact me at the contact email address on the website.

ISBN: **978-1-935881-21-6**

Design and layout: Jack Herris
Color Profiles: Bob Pearson
Cover design: Aaron Weaver
Digital photo editing: Aaron Weaver & Jack Herris

Aeronaut Books

www.aeronautbooks.com

Table of Contents

On the Cover

Steve Anderson's evocative painting of the famous Rumpler C.IV, the aircraft for which Rumpler is still known today. C.IV 8424/16 *Dalila* was from the first C.IV production batch. Only the highest-performance Allied fighters had any hope of intercepting a Rumpler C.IV if it was on a high-altitude reconnaissance mission. At high altitude the aerodynamic Rumpler C.IV was as fast as Allied fighters and had a good chance of besting them in combat.

Above: The Rumpler company test pilots in 1917; Gustav Basser at left.

Above: Well-known propaganda photo of a Rumpler C.I (upper right) flying over the Pyramids in Egypt. The photo of the Pyramids was actually taken by a Rumpler C.I but it could not take a photo of itself, so the C.I in the photo was added for propoganda value.

Rumpler Aircraft Designations

The Rumpler factory used alphanumeric designations for their aircraft designs. The first digit indicated the year of the design; such as "4" for 1914. A letter then indicated the category of the design, and the following number indicated the progression of designs in that category. So 5A2 would indicate the second design in the "A" category of designs started in 1915.

Military designations were entirely independent of the factory design designations. To help the reader navigate the designs, production aircraft are identified by their military designations with factory designations also given; prototypes were not given military designations and are identified by their factory designation.

Introduction

Above: Fitted with the 260 hp Mercedes D.IVa six-cylinder engine, the graceful Rumpler C.IV was as fast as Allied fighters at its ceiling, where few fighters could reach it. It became the premier German long-range photo-reconnaissance airplane. Over-compressed engines designed for more power at higher altitude maintained its edge throughout the war. The Rumpler C.IV airframe was also used for the specialized *lichtbild* (photo) aircraft. This early-production Rumpler C.IV of *Flieger Abteilung* 19 is barely distinguishable from a late-production Rumpler C.III. Later production C.IV aircraft discarded the propeller spinner and gained speed due to reduced drag. The high-flying Rumpler C.IV is the best-known Rumpler aircraft of the war and probably the most important.

Edmund Rumpler (Jan. 4, 1872–Sept. 7, 1940), founder of Rumpler Luftfahrzeugbau in November 1908, was a Vienna-born Austrian citizen. He was a gifted automotive engineer with many technical contributions to that industry before he had entered aviation due to its promising potential, becoming the first aircraft manufacturer in Germany.

On July 21 1910, Edmund Rumpler obtained an exclusive, five-year manufacturing license for *Taube* aircraft from Austrian inventor Igo Etrich. After a series of flight demonstrations, on Oct. 30, 1910, the Prussian army ordered five *Taubes* from Rumpler. However, the German patent office invalided Etrich's *Taube* patent in September 1911 and the *Taube* design became public domain. Immediately, many German companies started building aircraft to the *Taube* design due to its inherent stability and safety. Understandably, Rumpler then refused to honor the license agreement or pay license fees. Rumpler continued to build *Taubes* that were essentially the same as the Etrich design, and so Etrich successfully sued Rumpler. However, the amount of the damages awarded was small compared to the amount Rumpler was making from selling aircraft to the *Fliegertruppe*. The military was keen on the *Taube* due to its stability and safety, key advantages in these early days of aviation. As a result, the army purchased a wide variety of *Taubes* from different manufacturers; however, the greatest number were purchased from Rumpler. Rumpler *Taubes* set a number of records pre-war and gave Rumpler a good reputation.

The limitations of the *Taube* configuration were

Frontbeststand Inventory of All Rumpler Aircraft at the Front

Manufacturer	Type	1914 31 Aug	1914 31 Oct	1914 31 Dec	1915 28 Feb	1915 30 Apr	1915 30 Jun	1915 31 Aug	1915 31 Oct	1915 31 Dec	1916 28 Feb	1916 30 Apr	1916 30 Jun	1916 31 Aug	1916 31 Oct	1916 31 Dec	1917 28 Feb	1917 30 Apr	1917 30 Jun	1917 31 Aug	1917 31 Oct	1917 31 Dec	1918 28 Feb	1918 30 Apr	1918 30 Jun	1918 31 Aug
Rumpler	A/13	2																								
	A/14	8	3																							
Rumpler	B.I	1	6	21	32	41	38	17	14	6	5	4	2		8											
Rumpler	C.I									29	67	120	144	202	231	190	145	107	55	37	40	16	11	9	5	4
	C.Ia												1		2	1	1	52	162	129	30	5	3	3	2	1
	C.III															1	42	22	15	7	1	1	1	1		
	C.IV																	24	122	257	225	208	161	187	182	110
	C.V																				3	9			4	6
	C.VII																					17	48	94	112	85
	C.IX																					20	13	11	5	2
Rumpler	Rubild																				11	64	51	45	39	99
	Rubild Mb																							50	43	78
Rumpler	G.I												1		1											
	G.II												1	7	8	4	1	1								
	G.III															1		3	5	5	10	1	4			

Above: The *Frontbeststand* table above summarizes all Rumpler Army aircraft of all types at the front during the war. The Rumpler A-types were *Taube* designs quickly replaced by the superior B-type unarmed biplanes. Rumpler was especially successful with C-type armed two-seaters and *Rubild* aircraft, which were specialized high-altitude reconnaissance airplanes with strip cameras. No other manufacturer built *Rubild* aircraft. The Rumpler C.IV, C.V, C.VII, C.IX, and *Rubild* aircraft all used the same airframe; these types were distinguished by different engines and equipment. Rumpler bombers served in small numbers for an extended period. Naval aircraft are not included in the *Frontbeststand* tables.

SVK Table of Rumpler Seaplane Orders and Deliveries

Order Number	Type	Marine Numbers	Design	Class & Engine		1914 J	F	M	A	M	J	J	A	S	O	N	D	1915 J	F	M	A	M	J	J	A	S	O	N	D
1		4	E4		100Arg	} Order & delivery not determined																							
2		8	E8																										
3		47	DD		150B										Ordered again					(1) Rejected delivery									
4		49–51	DD		150B							(3)	2	1															
5		64	DD		100M											(1)	Delivery not determined. Cancelled 6-6-1												
6		86–90	4B11	B	100M								(5)			2	2	No.89 not delivered											
7		101–110	4B12	B	150B												(10)												4
8		241–253	4B12	B	150B							(13)									3	8			1				
9		436–440	4B12	B	150B												(5)						1						
10		441–442	4B12	B	150B									(2)										2					
11	751	751	6B1	ED	160M																								
12	751	787–788	6B1	ED	160M																								
13	751	890–899	6B1	ED	160M																								
14	751	1037–1046	6B1	ED	160M																								
15	751	1047–1061	6B1	ED	160M																								
16	1064	1062–1066	6B2	ED	160M																								
17	1064	1188–1207	6B2	ED	160M																								
18	1064	1434–1458	6B2	ED	160M																								
Orders that Month												16	5	2		1	15			1									
Deliveries that Month													2	1		2	2				3	8	1	2	1				4
Orders that Year																	39												1
Deliveries that Year																	7												19

Above: The Rumpler factory at Johannisthal in 1914.

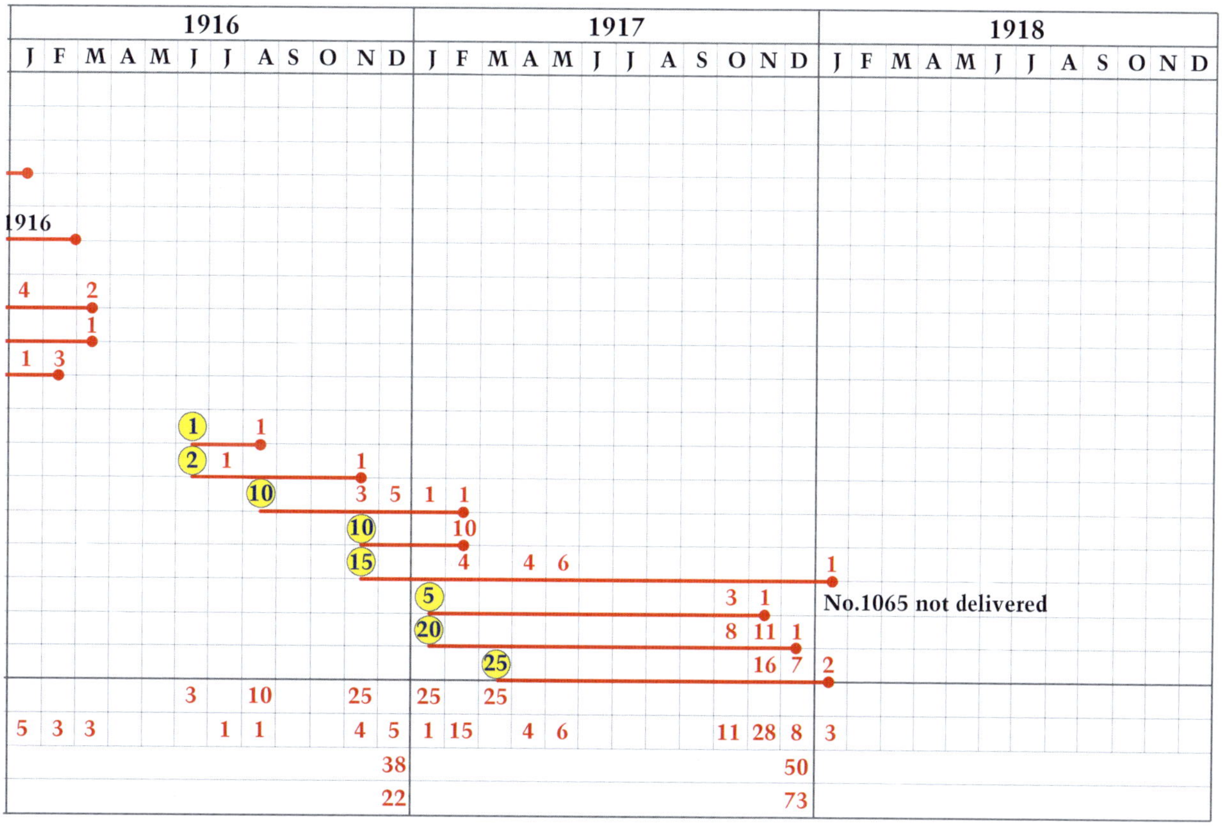

soon evident and, like other manufacturers, Rumpler soon moved on to biplanes, building a number of related landplane and seaplane designs that were built in reasonable quantity for the time. From there Rumpler branched into several directions, including armed reconnaissance airplanes and bombers. The first all-Rumpler design to make a significant impact was the Rumpler C.I general-purpose armed two-seater. The C.I was an excellent design that reached the front in late 1915 and was used throughout the rest of the war, at first operationally and later built as a trainer due to its excellent handling qualities. Rumpler built several bomber designs in small quantities, but the C.III and C.IV high-altitude reconnaissance airplanes that followed the C.I were the most successful and famous Rumpler designs.

In early 1916 Rumpler started development of two advanced designs, a high-altitude fighter and

Left: Rumpler *Taube* in production. Rumpler gained an early reputation based on building and selling copies of the Etrich *Taube*. Rumpler *Taubes* won many aviation awards before the war.

Below: The Rumpler military flying school at Müncheberg circa 1916/17. Rumpler B.I trainers crowd the hangar.

a related high-altitude, two-seat reconnaissance airplane. The reconnaissance airplane development was abandoned after two prototypes were tested but Rumpler persevered with development of the fighter, eventually designated the Rumpler D.I, for the duration of the war.

Rumpler was one of the most innovative aircraft manufacturers and continually worked to improve aerodynamic performance of their aircraft. Rumpler also did much work on advanced structural designs in conjunction with their aerodynamic advances.

After the war the German aviation industry was targeted for destruction by the victorious Allies and Rumpler returned to the automotive industry. His *Tropfen-Auto*, revealed in 1921, was a remarkable design for its time with a coefficient of drag (measured in 1979) of only 0.28, better than many automobiles currently on sale.

Above: Buiilt from 1921 until 1925, Rumpler's *Tropfen-Auto* was highly streamlined, with a drag coefficient of only 0.28, better than many modern automobiles. Rumpler specialized in excellent engineering and highly streamlined designs. (Photo courtesy Detlef Garbrecht)

Above: Edmund Rumpler was a highly-accomplished automotive engineer before founding his aircraft manufacturing company. He designed the first German auto engine to combine the engine and transmission in one unit, and patented a swing-axle rear suspension later adopted by Ferdinand Porsche for the VW Beetle and Porsche 356. With German aviation effectively strangled by the Armistice terms, after the war Rumpler returned to the auto industry.

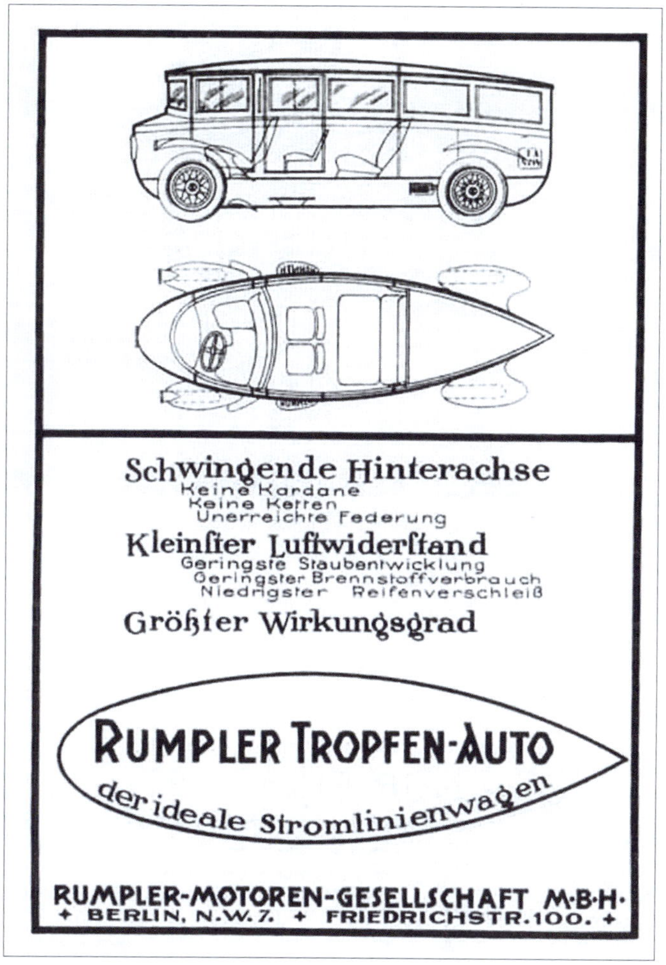

Rumpler Taubes

Above: The Rumpler 3C *Taube* was more streamlined development of the original Rumpler *Taube*, itself a copy of the Etrich *Taube*. Power was a 100 hp Mercedes D.I six-cylinder engine. The large wing cut-outs are to give the pilot and observer good downward visibility for its role as a reconnaissance aircraft. A claw brake is fitted under the axle.

On July 21 1910, Edmund Rumpler obtained an exclusive, five-year manufacturing license for *Taube* aircraft from Austrian inventor Igo Etrich. After a series of flight demonstrations, on Oct. 30, 1910, the Prussian army ordered five *Taubes* from Rumpler. However, the German patent office invalided Etrich's *Taube* patent in September 1911 and the *Taube* design became public domain. Immediately, many German companies started building aircraft to the *Taube* configuration due to its inherent stability and safety. Understandably, Rumpler then refused to honor the license agreement or pay license fees. Rumpler continued to build *Taubes* that were essentially the same as the Etrich design, and so Etrich successfully sued Rumpler. However, the amount of the damages awarded was small compared to the amount Rumpler was making from selling aircraft to the *Fliegertruppe*. The military was keen on the *Taube* due to its inherent stability and safety, a key advantage in these early days of aviation. As a result, the army purchased a wide variety of *Taubes* from different manufacturers, with the greatest number being purchased from Rumpler.

Above: Streamlining became a Rumpler hallmark. The Rumpler 3C *Taube* was the aircraft type in which Linnekogel climbed out of the cockpit in flight and held on to the support pylon while the aircraft flew itself hands-off. This demonstrated not only his courage but the inherent stability of the Taube design that was so prized before the war when primitive aviation was very hazardous.

Rumpler Taubes Purchased by the Fliegertruppe (Prussian Army)

Year	Quantity
1911	10
1912	48
1913	73
1914	3

The *Taube's* limitations gradually became evident, and few were purchased after 1914, when the configuration was clearly obsolete. The *Taube* had high drag, so speed, climb, and ceiling were limited. And in combat, its inherent stability became a problem; it was simply too stable and difficult to maneuver to dodge anti-aircraft fire or elude an attacking aircraft. Fortunately, there were few attacking aircraft during the war's early months.

On September 4, 1913, a Rumpler *Taube* lost a wing during the annual army maneuvers and the two airmen were killed. This caused the German military authorities to perform their first serious structural tests on aircraft, with the result that the Rumpler *Taube's* safety factor was found to be 3.42 at best. To express it differently, the Rumpler *Taube* could withstand a G-loading of 3.42 G's before its structure failed. To put that in perspective, a normal category light aircraft in the USA must sustain a load factor of 3.8 without *damage*, and 150% of that stress, or a load factor of 5.7 Gs, before structural failure occurs. To be certified as an acrobatic aircraft, the load factors would be 6.0 Gs without damage and 9.0 Gs before structural failure. To return to the Rumpler *Taube*, one aircraft (A.28/12) tested had a safety factor of 3.42, but another, A.46/13, had a safety factor of only 2.89! This was not strong enough to be safe for routine flight let alone the stress of combat missions.

Recognizing the *Taube* configuration was obsolete, the army started purchasing biplanes before the war started and disposed of 55 older, worn out *Taubes* the end of June 1914, just before the war. Although the army had to purchase additional *Taubes* early in the war to make up losses, the *Taube's* days were clearly over. Interestingly, Rumpler *Taubes* did not last long at the front after hostilities started; the Gotha *Taubes* were more robust and better able to withstand the rigors of operational service.

Above & Left: Early Rumpler *Taubes* displaying the characteristic *Taube* wing and tail planforms.

Below: This Rumpler *Taube* crashed on July 20, 1914 , proving that inherent stability is not everything.

Below: Rumpler *Taube* A.10/12 was an early, pre-war aircraft with a four-cylinder Argus engine. The observer was seated in front for best visibility, and the aft-seated pilot had wing cut-outs for enhanced downward field of view.

Right: Rumpler *Taube* of *Flieger-Abteilung* 35.

Below: Early Rumpler *Taube*.

Below Right: Rumpler *Taube* in the snow is visited by a woman; details are not known. The national insignia appears to be pained on the elevator in addition to the normal locations.

Below: In September 1914 aerial reconnaissance by both airplanes and airships was critical to German victories over much larger Russian forces at Tannenburg. In recognition of this historic feat *General* von Francois is photographed in front of the Rumpler *Taube* of *Lt*. Canter, one of the heroic flyers of Tannenburg, and his observer at right.

Above: The Rumpler *Taube* 4C was the final Rumpler *Taube* design. For improved maneuverability it had conventional hinged ailerons and elevators; these were much easier for the pilot to operate. (The Peter M. Bowers Collection/ The Museum of Flight)

Above: The tail of the Rumpler 4C looked like that of the Rumpler B.I biplane that was soon to follow the *Taube* into production.

Right: The Rumpler *Taube* 4C was usually powered by a 100 hp Mercedes D.I engine, although a 125 hp Benz Bz.II was also used.

Right: The Navy used land-based Rumpler *Taube* in addition to Rumpler *Taube* floatplanes. Rumpler *Taube* given Naval serials include S8, S9, S 20, S21, S22, S42 (formerly A.131/13), and S43 (formerly A.133/13), the last two transferred from the Army.

Left: Early Rumpler *Taube* with very promient national insignia. Soldiers were likely to fire on any aircraft and aircrew of both sides frequently experienced "friendly fire"' on that account. Airmen know that "friendly fire" isn't friendly.

Right: Postcard of a flyer in front of an early Rumpler *Taube*.

Right: Crewmen in front of an early Rumpler *Taube* with "Rumpler Eindecker" painted on the fuselage and tactical number "16" on the aft fuselage.

Above: The Rumpler *Taube* 4C was usually powered by a 100 hp Mercedes D.I. (The Peter M. Bowers Collection/ The Museum of Flight)

Above: This Rumpler *Taube* 4C has drawn a crowd at an airshow.

Right: Rumpler *Taube* 4C with eye-catching lettering identifying the manufacturer. It worked; Rumpler was one of the most well-known pre-war German aircraft manufacturers.

Early Rumpler Seaplanes

Above: The Rumpler 4B12 was powered by a 150 hp Benz Bz.III. The Rumpler 4B series were developments of the 4A13 landplane. This unmarked aircraft may have been the prototype seaplane; it still uses the small rudder of its landplane ancestors while later seaplanes had a larger rudder due to the destabilizing effect of the floats.

Starting in 1914 Rumpler developed a flying boat that remained a single prototype and a small series of floatplanes developed from the Rumpler 4A13, itself a more powerful development of the Rumpler B.I unarmed two-seat trainer and reconnaissance airplane.

As was typical of German naval aircraft. there were a number of variations on the basic design with limited production. As the war progressed Rumpler focused on two-seat reconnaissance airplanes for the Army, which purchased almost 20 times more airplanes than the Navy.

Rumpler 4B11, 4B12, & 4B13

The Rumpler 4B11 and 4B12 were two-seat reconnaissance and training floatplanes based on the Rumpler 4A13 landplane. Both were built in very small numbers for the Navy. The 4B11 was powered by a 100 hp Mercedes D.I. The 4B12 had the more powerful 150 hp Benz Bz.III. The 4B13 was similar except powered by a 160 hp Gnome rotary engine. All three types were conventional wood, wire, and fabric designs typical of their era.

Rumpler 4B Production		
Type	Qty	Marine Numbers
4B11	7	49, 64, 86–90
4B12	31	51, 101–110, 241–253, 436–442
4B13	1	50

Above: Rumpler 4B11 Marine Number 49.

Below: Rumpler 4B12, Marine Number not known.

Right: Rumpler 4B12 Marine Number 108.

Above: Rumpler 4B12 Marine Number 250. (The Peter M. Bowers Collection/The Museum of Flight)

Left: Rumpler 4B12 Marine Number 436.

Above: Rumpler 4B12; no Marine Number is visible and the rudder is smaller than production machines, indicating this is the 4B12 prototype.

Left: Rumpler 4B12 Marine Number 252 has yet another fin and rudder design.

Below: Rumpler 4B12 apparently damaged in a landing accident.

Right: Rumpler 4B12 in flight.

Below: Rumpler 4B12 serves as a background for a crew portrait.

Below: Rumpler 4B12, Marine Number 105, and an FF29 at either Flying Station List or Sylt.

Rumpler 4E Flying Boat

The Rumpler 4E was a flying boat built in 1914 powered by a 150 hp Benz Bz.III engine. Records indicate that only one aircraft, Marine Number 47, was built.

This Page: The Rumpler 4E flying boat seemed to be much more popular with postcard publishers than with the Navy; there are three postcards of it here and a fourth on the next page.

This Page: Like most Rumpler designs the Rumpler 4E was well thought out and constructed, with clean lines for a biplane flying boat. Regardless of its qualities, the German Navy favored floatplanes over flying boats due to the cold waters around Germany. Floatplanes kept the crew further away from the cold water and kept them dryer and warmer.

Rumpler Reconnaissance Airplanes

Reconnaissance was the most important role of aviation in the Great War, and Rumpler's greatest contribution was its line of excellent reconnaissance airplanes that served at the front for the entire war. The pre-war Rumpler *Taubes* served during the early months of the war but by the end of 1914 had been withdrawn from the front. In their place flew the Rumpler B.I, followed in late 1915 by the excellent C.I. The transitional C.III of late 1916 led to the exceptional Rumpler C.IV family, which provided Germany with high-altitude, long-range reconnaissance until the Armistice that was seldom hindered by Allied fighter interception.

Rumpler B.I

Above: The Rumpler B.I was of conventional wood, wire, and fabric construction and used a 100 hp Mercedes D.I engine.

By 1914 it was increasingly clear that the biplane had an inherently superior performance compared to the *Taube*, having lower drag and therefore higher speed, climb rate, and ceiling. Rumpler designed their unarmed biplane, factory designation 4A, in 1914, and the aircraft was purchased by the army as the Rumpler B.I. Powered by a 100 hp Mercedes D.I, the Rumpler B.I was a good aircraft for its time, with good performance and flying qualities, and gave satisfactory service. The Rumpler 4A13 derived from the 4A/B.I had a balanced rudder and the same engine. The Rumpler 4A14 was similar but powered by a 150 hp Benz Bz.III.

Rumpler B.I Specifications

Engine:	100 hp Mercedes D.I	
Wing:	Span	13.0 m
General:	Length	8.4 m
	Height	3.1 m
	Empty Weight	750 kg
	Loaded Weight	970 kg
Maximum Speed:		145 kmh

The Austro-Hungarian *Luftfahrtruppen* purchased a total of 31 Rumpler B.Is. They were used by *Fliks* 4, 5, 9, and 13 for bombing and reconnaissance and were used as trainers at least through August 1918.

Above & Right: This aircraft may be the Rumpler B.I prototype. Above is Sanke card 260 and at right is a slightly different view. The B.I was a reliable replacement for the obsolete *Taube* and had good flying qualities, making it suitable for training use as well as operational reconnaissance missions.

Left: Rumpler B.I in left foreground in a hangar with other B-types.

Above: Rumpler B.I depicted in Sanke card 271. Aircraft recognition, always a problem for the infantry, was in its infancy and the aircraft is will decorated with German national insignia to identify it for the ground troops.

Below: Rumpler B.I B.362/15.

Above: Rumpler B.I B.483/14.

Above: Naval Rumpler B.I S60.

Above, Below, Above Right, & Below Right: Anonymous Rumpler B.I aircraft in use.

Right: Rumpler B.I approaching to land. In addition to the *Fliegertruppe*, the B.I was used by the German Navy and 31 were sold to the Austro-Hungarian *Luftfahrtruppen* where they served until at least August 1918.

Above: Rumpler 4A14 was developed from the Rumpler B.I by installing a 150 hp Benz Bz.III.

Below, Below Right, Bottom Right: Rumpler B.I aircraft.

Above: Rumpler B.I with then *Vzfw*. Raetsch at right.

Left: Rumpler B.I from the *Militär Flieger Schule* Müncheberg flown by *Lt*. Raetsch on its nose.

Military aircraft serials for 1914 and 1915 are complicated, as there were impressments and aircraft taken over from the factories as available. This means the serials were often not typical numbers. Sometimes the aircraft type changes with each serial within a series of numbers. (Data courtesy Reinhard Zankl.)

Rumpler B.I

1914

B.362/14 – B.395/14	(34 a/c)	Maybe more, but certainly not more than B.360/16 to B.406/16
B.470/14 – B.487/14	(18 a/c)	
B.509/14	(1 a/c)	Range 506–511/14 so far uncovered by other types*
B.567/14	(1 a/c)	Range 562–570/14 so far uncovered by other types*
B.669/14	(1 a/c)	Also noted as Av B.I
B.708/14 – B. 760/14	(53 a/c)	Bordered by Av. B.I 707/14 and Alb B.II B.766/14*
B.959/14	(1 a/c)	Bordered by LVG B.I 950/14 and Go B.I 960/14*
B.1018/14 – B.1042/14	(25 a/c)	Bordered by Alb B.I 1017/14 and Fok B.I 1043/14
B.1196/14	(1 a/c)	

1915

B.352/15 – B.355/15	(4 a/c)	Bordered by Eu B.I 350/15 and Av B.I 359/15*
B.367/15	(1 a/c)	May be a mis-recorded B.367/14
B.383/15	(1 a/c)	
B.955/15 – B.966/15	(12 a/c)	

Note: * doesn't mean the uncovered serials have to be Rumpler B.I

Several Rumpler B.I aircraft were used by the Rumpler flying school (*Militär Flieger Schule* Müncheberg) which were owned by the Rumpler factory and had no military serial.

The Navy received 24 Rumpler B.I aircraft with various engines:

S.37 – S.41	(5 a/c)	S.128 – S.135	(8 a/c)
S.59 – S.65	(7 a/c)	S.163	(1 a/c)
S.93 – S.95	(3 a/c)		

Rumpler C.I & C.Ia

Above: Early production models of the Rumpler C.I were powered by the 160 hp Mercedes D.III, and immediately established themselves as reliable airplanes with good performance and handling qualities. Some later production C.Is had a 150 hp Benz Bz.III or 180 hp Argus As.III. The C.I was an excellent general purpose C-type that was popular with its crews.

The Rumpler C.I was an armed, two-seat general purpose reconnaissance airplane designed based on the lessons learned from the very first armed two-seaters from Albatros, Aviatik, and LVG, all of which were derivatives of earlier, unarmed two-seaters. Designed to carry guns, bombs, and wireless, the Rumpler C.I had good speed and maneuverability and very good climb. Coupled with its robust construction, its good performance and versatility quickly earned it a good reputation among the aircrew who flew it.

When it first flew in the summer of 1915, the Rumpler C.I was faster than the Fokker and Pfalz *Eindecker* fighters and had an equivalent climb rate. *Idflieg* ordered the first batch of aircraft in July 1915 and the C.I passed military acceptance testing in October. The first C.I aircraft arrived at the front in November–December 1915. Armed with both a fixed and a flexible machine gun and with performance exceeding the *Eindeckers*, the Rumpler C.I was an excellent combat aircraft for its time and lasted a long time at the front.

The Rumpler C.I had a wood and fabric airframe typical of the time and was powered by a 160 hp Mercedes D.III or a 150 hp Benz Bz.III. Hannover built the Rumpler C.Ia powered by the 180 hp Argus As.III; speed at low levels was improved slightly at the expense of reliability and high-altitude performance. The C.Ia was used primarily on the Eastern Front, where they were not generally faced with the latest fighter opposition, and gave good service.

Unlike similar aircraft, the C.I had a steel-tube framework integrated with the wood structure of the forward fuselage, making it stronger and more survivable in case of a crash. The robust, versatile C.I was used for a variety of experimental programs testing new wings and engines. The large, 220 hp Mercedes D.IV straight-eight was tested in the C.I, as was the first BMW.IIIa engine.

Rumpler C.I aircraft built under license for training were powered by a variety of engines. Aircraft built for training service seldom used the 160 hp Mercedes D.III because it was needed for

Rumpler C.I Specifications		
Engine:	160 hp Mercedes D.III	
Wing:	Span (Upper)	12.15 m
	Span (Lower)	10.04 m
	Chord (Upper)	1.75 m
	Chord (Lower)	1.75 m
	Gap	1.86 m
	Area	35.7 m^2
General:	Length	7.85 m
	Empty Weight	793 kg
	Loaded Weight	1,333 kg
Maximum Speed:		152 kmh
Climb:	3000m	25 min

fighters. Re-built 150 hp Benz Bz.III engines were installed along with the 175 hp Rapp Rp.IIIa, 180 hp Argus As.III, and 185 hp Conrad C.III(Nag). Trainers with the Conrad engine were built by Germania and designated C.Ic(Germ). As far as is known, the only license-built Rumpler C.I aircraft used in combat was the Hannover-built C.Ia. Manufacturers who built the C.I for training use were Märkische, Germania, Hansa-Brandenburg, the Bayerische Rumpler-Werke (owned by Rumpler), and the Flugzeugwerke Albert Rinne. Details are in the production table at right.

Finally, it should be noted that the C.I airframe had such good qualities that it was used for the successful Rumpler 6B1 single-seat floatplane fighter. Despite being derived from a two-seat reconnaissance airplane, the 6B1 was as successful as the competing single-seat floatplane fighters derived from land-plane fighters.

Above: This Rumpler C.I 1851/15 came down behind Allied lines and was captured.

Rumpler C.I Production Orders

Date	Qty	Serials	Notes
Jul. 1915	50	393–442/15	
1915	4	1025–1066/15	
1915	20	1588–1599/15	Other serials unk.
Oct. 1915	51	1836–1886/15	
Dec. 1915	150	4515–4664/15	
Mar. 1916	50	1125–1174/16	
Mar. 1916	100	2600–2699/16	
Sep. 1916	50		Cancelled
Sep. 1916	200	4600–4699/16	Han. C.Ia
Oct. 1916	100	5400–5499/16	Mark.; trainers
Oct. 1916	100	6075–6174/16	Germ.; trainers
Oct. 1916	75	6500–6574/16	HaBra.; trainers
Jan. 1917	75	900–974/17	Han. C.Ia
Mar. 1917	200	2550–2749/17	Mark.; trainers
Apr. 1917	100	4100–4199/17	Ha-Bra.; trainers
May 1917	100	4900–4999/17	Germ.; trainers
May 1917	100	5475–5574/17	Han. C.Ia
Jul. 1917	200	2550–2749/17	Rinne; trainers
Sep. 1917	250	See Note 1	Mark.; trainers
Oct. 1917	100	14100–14199/17	Germ.; C.Ic trnrs.
Apr. 1918	150	3000–3149/18	Mark.; trainers
Jun. 1918	100	7050–7149/18	Bayru.; trainers
Jul. 1918	100	See Note 2	Li-Ho; cancelled
Oct. 1918	50	2550–2749/17	Germ.; trainers

Notes: 1. Known serials are 12820–13069/17.
2. Serial C.13074/18 from this order is known.
3. 2363 Rumpler C.I aircraft built; 463 by Rumpler, 375 by Hannover (Han.), 700 by Märkishe (Mark.), 350 by Germania (Germ.), 175 by Hansa-Brandenburg (HaBra.), 100 by Bavarian Rumpler Works (Bayru.), 200 by Flugzeugwerke Rinne.

Right: Rumpler C.I with two 50 kg PwK bombs carried under the fuselage.

Above & Below: The Rumpler 5A1 was the prototype Rumpler C.I. The slightly modified production Rumpler C.I had the factory designation Rumpler model 5A2. The rudder and fin of the production aircraft was enlarged slightly from the 5A1. The aircraft is seen in its initial form without markings at the time of initial test flights in June 1915 in front of the Rumpler factory at Johannisthal.

Above & Below: The Rumpler C.I prototype after national insignia were applied but before the pilot's fixed, synchronized gun was fitted. A spherical gravity tank has been added below the center section. The gunner is demonstrating his wide field of fire with a dummy wooden gun.

Above & Below: The Rumpler C.I prototype before the pilot's fixed, synchronized gun was fitted. The spherical gravity tank added below the center section is clearly visible. The claw brake is fitted to the axle.

This Page: Rumpler C.I 1851/15 was from the fourth production batch. These were the first Rumpler C.I aircraft built with a rotating turret for the gunner, something that all subsequent aircraft featured.

Above: Rumpler C.I C.2642/15

Top: Rumpler C.I 1869/15 draws a crowd.

Left: Rumpler C.I observer demonstrates his weapon.

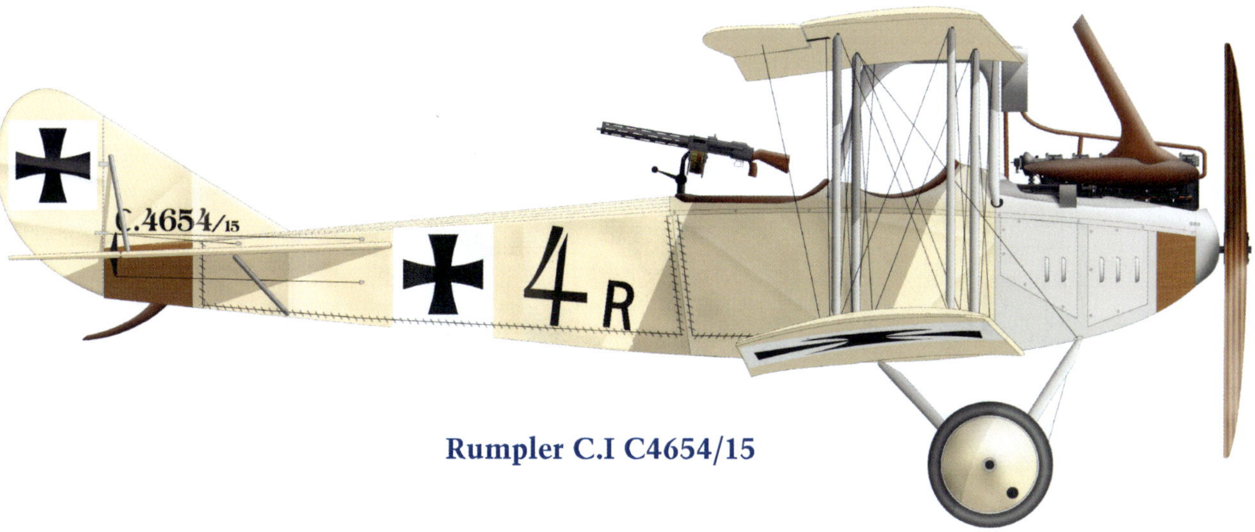

Rumpler C.I C4654/15

Rumpler C.I

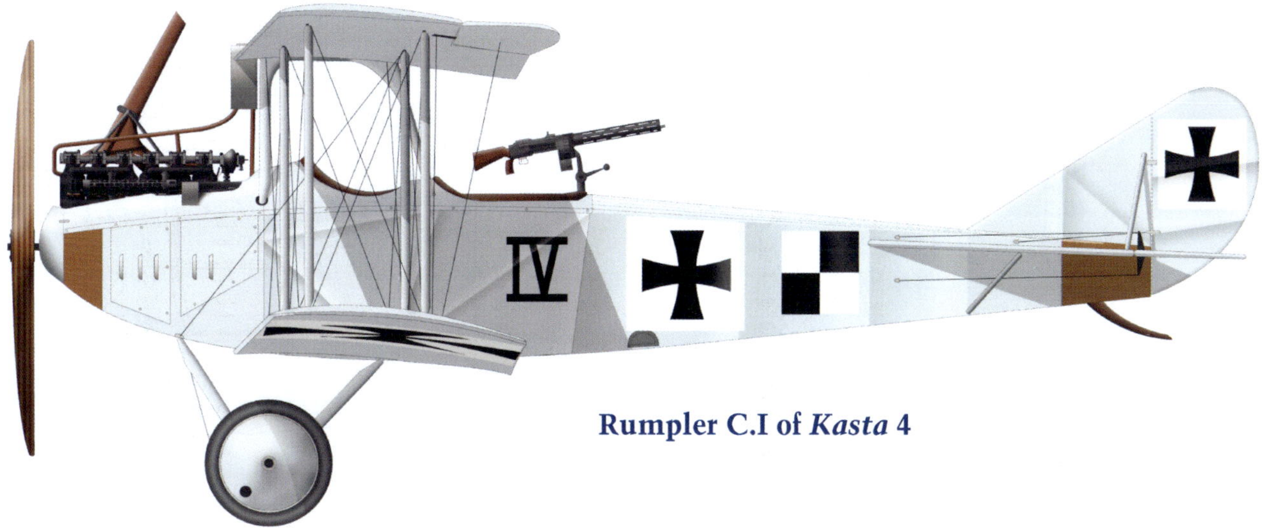

Rumpler C.I of *Kasta* 4

Above: The desire to create 'low observable' or 'stealthy' aircraft difficult to detect is not new. Here a Rumpler C.I has been covered by transparent Cellon in an experiment to make it more difficult to see. Other types involved in these trials included the Fokker E.III and Linke-Hofmann R.I. However, Cellon proved to have insurmountable disadvantages. First, it was shiny, making the aircraft *more* visible in situations where bright sunlight illuminated it. Second, Cellon was brittle and tore easily, and it sagged when it absorbed water, ruining the aircraft's aerodynamic qualities. An interesting series of experiments, Cellon covering was quickly abandoned.

Above: Rumpler C.I 53/15 was a very early production aircraft.

Above: Rumpler C.I 4601/15 in pristine finish. The finish is not original so it has been recovered, probably after repair.

Above: Rumpler C.Ia(Han) 4645/16 after a landing incident. It has the distinctive built-up decking for the gun ring and camouflage of Hannover-built aircraft.

Above: Rumpler C.I 4658/15 after a landing incident.

Below: Rumpler C.I 1131/16 after being captured.

Below: Rumpler C.I 6162/16 trainer after a bad landing.

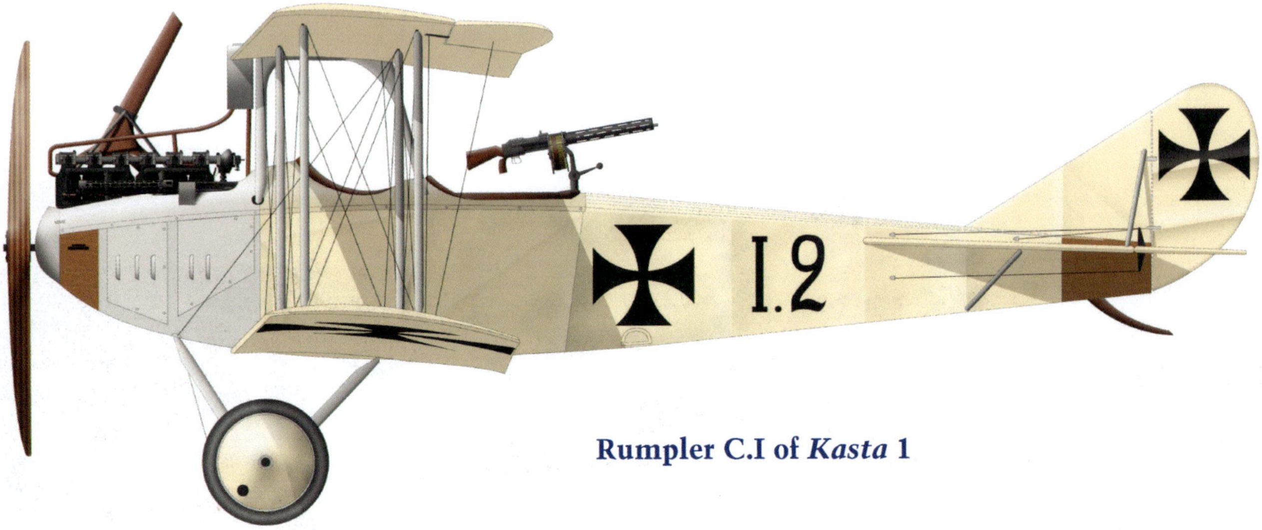

Rumpler C.I of *Kasta* 1

Rumpler C.I of *Kasta* 1

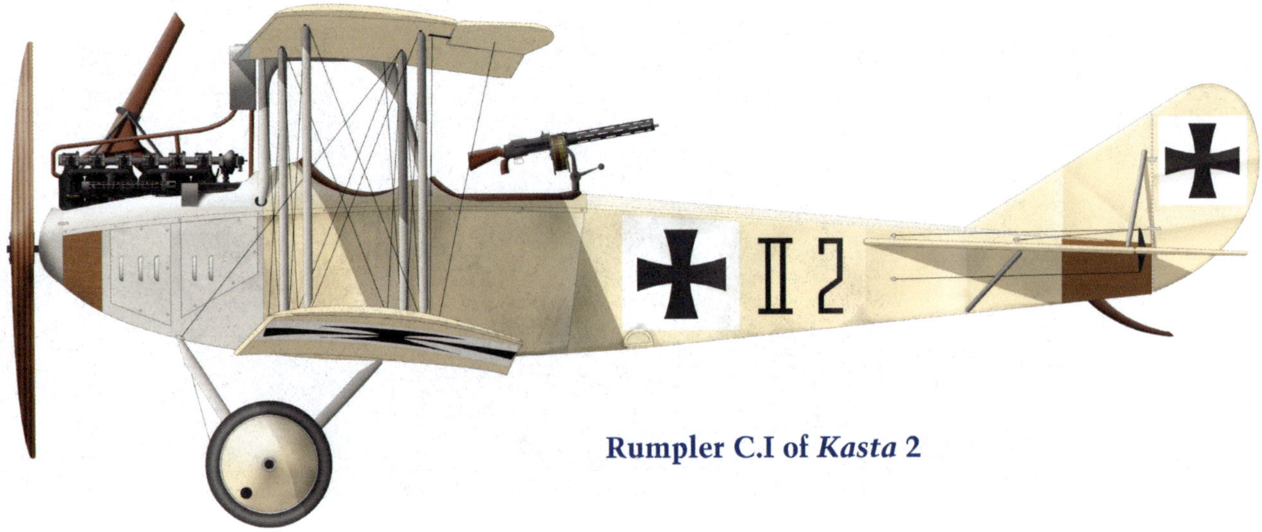

Rumpler C.I of *Kasta* 2

Above: Rumpler C.I serving with *Feld-Flieger-Abteilung* 32.

Above: Rumpler C.I 4660/16 was immortalized in Sanke card 439.

This Page & Facing Page: The Rumpler C.I was a popular aircraft and is certainly a candidate for having the most air-to-air photographs taken of it of any WWI German airplane.

This Page: More of the many air-to-air photographs taken of the Rumpler C.I.

Right: Rumpler C.I trainer, under the wing is a 1918 cross.

Above: Rumpler C.I cruises serenely above the clouds.

Left: Rumpler C.I 1131/16 in French hands.

Below Left: Rumpler C.I in Allied hands after capture.

Below: Rumpler C.I ready to take off on another mission.

Above: Rumpler C.I 393/15 was the very first production Rumpler C.I.

Right: This Rumpler C.I was retired from flight duties and converted into a ground-based gunnery trainer. It is mounted on a complex platform allowing it to move for greater realism while the gunner aims at targets pulled past the trainer. A target, the silhouette of a Rumpler, is visible in the background.

Rumpler C.I of *Schusta* 2

Above: This closeup of a Rumpler C.I with cowling panel removed shows the pilot's fixed machine gun. The "H" with arrow added to the photo points to the '*Hebel*', the lever arm of the synchronizing mechanism.

Right: Rumpler C.I ready for its next flight.

Below: Camouflaged Rumpler C.I assigned to *Schusta* 2.

Left: Rumpler C.Ia(Han) 4666/16 was from the first production batch of C.Ia aircraft built by Hannover. The C.Ia was powered by a 180 hp Argus As.III, giving it slightly greater speed at low level but poorer reliability and high-altitude performance than the 160 hp Mercedes D.III. All C.Ia aircraft were built by Hannover, the only C.I derivative Hannover built.

Above: Rumpler C.Ia(Han) with the raised fuselage around the observer's gun ring unique to Hannover-built aircraft. The only license-built C.I aircraft used in combat were the C.Ia aircraft built by Hannover. All other license-built aircraft were intended as trainers.

Left: Rumpler C.I under test with a 175 hp Rapp Rp.IIIa engine at Adlershof.

Above: Another view of the Rumpler C.I under test with a 175 hp Rapp Rp.IIIa engine at Adlershof.

Left: Rumpler C.Ia(Han) aircraft lined up outside the Hannover factory ready for delivery.

Below: Rumpler C.Ia(Han) at the Hannover factory.

Above: Rumpler C.Ia(Han) showing the distinctive built-up fuselage for the observer's gun ring unique to Hannover.

Above: Rumpler C.Ic(Germ) 13013/17 during a training flight with instructor Poelke in the rear cockpit.

Left: Rumpler C.I(Habra) trainer. The bulge in the lower port wing is the housing for the compass.

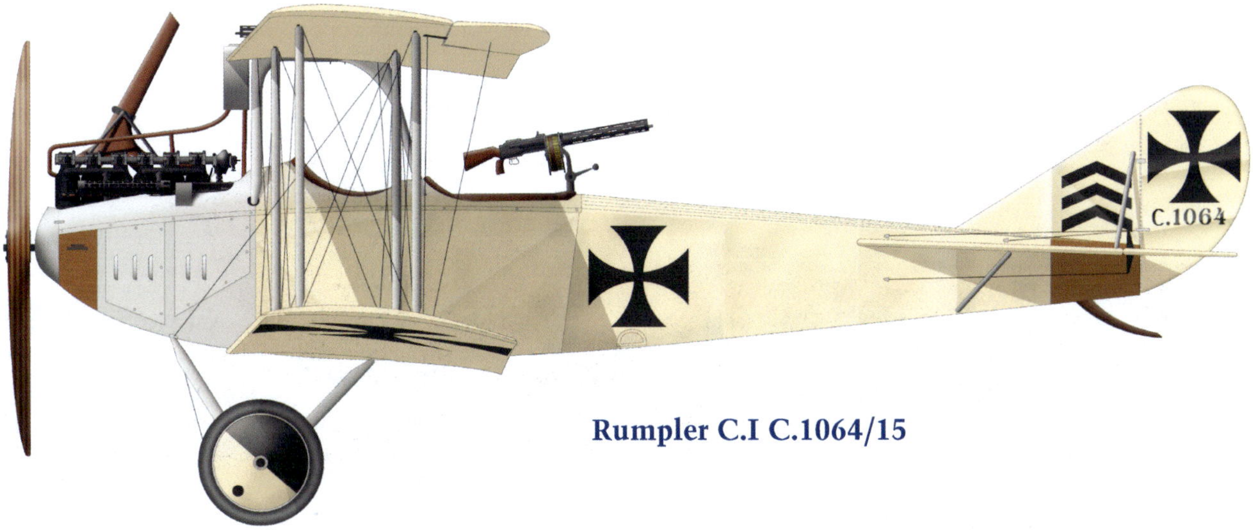

Rumpler C.I C.1064/15

Rumpler C.I C.4535/15 of *Kasta* 20

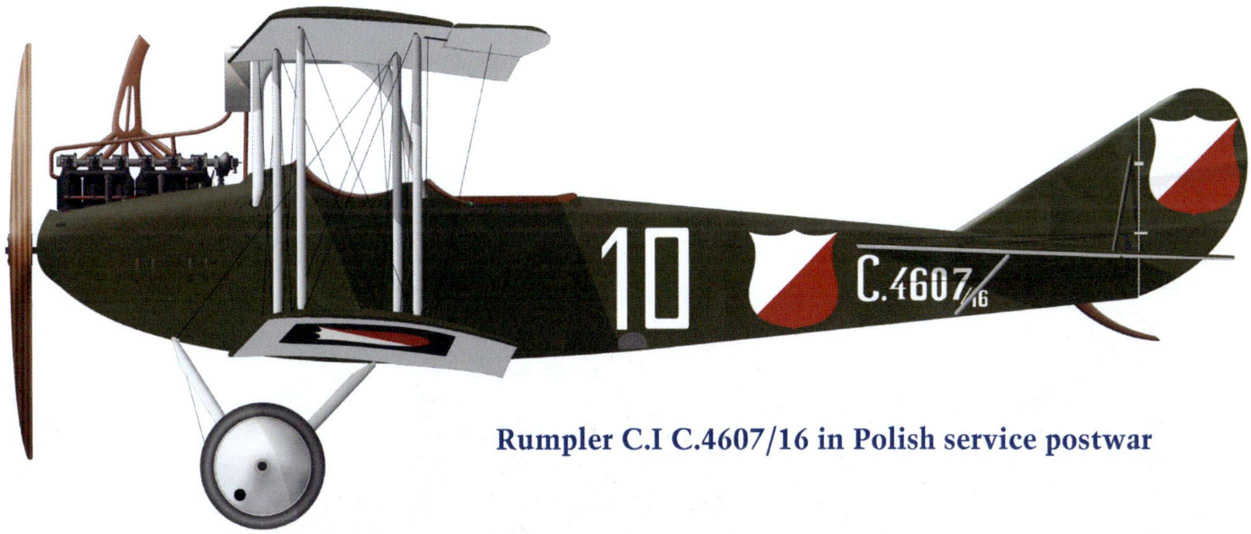

Rumpler C.I C.4607/16 in Polish service postwar

Above: Captured Rumpler C.I on display in France.

Above, Right, Below Right: Captured Rumpler C.I 4525/15 on display in France.

Above: Observer demonstrates his camera technique.

Left: Rumpler and its aircrew; von Hackenberger at left.

Right: Rumpler C.I(Bayru) 7051/18 was built as a trainer. Its late-style national insignia reflect the year 1918. Here is is undergoing acceptance flight testing at Adlershof in Autumn 1918.

Left: Rumpler C.I(Mark) 3042/18 was built as a trainer. Here is is undergoing acceptance flight testing at Adlershof in 1918.

Below & Below Left: The Rumpler C.I soldiered on postwar. Willy Donner and others pose with a C.I in transport service.

Below Right: Testing of the Rumpler 5A4 prototype powered by a 160 hp Mercedes D.III was abandoned after the third flight because engine and propeller vibration broke ribs in the lower wings.

Below: A Rumpler C.I in training service in 1918 (per the national insignia style) is a backdrop for a unit photograph.

Left & Below Left: The Rumpler C.I served at the front in quantity for a long time, and a number were downed behind Allied lines and captured. Two of these are shown here.

Left & Above: The Rumpler C.I served with the Latvian air service postwar.
The finish was German camouflage-printed fabric and the Latvian national markings consisted of a red Swastika over a white background.

Rumpler C.III

Above: A Rumpler C.III prototype is shown above. Although it looks very similar to the later and more well-known C.IV, it has a curved rudder without fixed fin, and had inadequate directional stability. The raised decking behind the observer's cockpit was soon replaced by a conventional design. All controls were aerodynamically balanced.

The Rumpler C.III (6A5) was a transitional production type between the better-known and more successful C.I and C.IV two-seaters. The Rumpler C.I, powered by the 160 hp Mercedes D.III engine, entered service in 1915 and was an immediate success. It was kept in production for the remainder of the war, initially as a combat aircraft, later as a trainer due to its excellent handling characteristics. Rumpler C.II was the designation reserved for a production version of the Rumpler 5A3 prototype, which was basically a modified C.I powered by a 180 hp Argus As.III engine. However, only one prototype was built and the C.II did not go into production. Further Rumpler two-seater development through a series of prototypes lead to the Rumpler C.III.

Whereas the C.I was a general-purpose observation plane, the C.III was intended as a higher performance aircraft for high-altitude photographic reconnaissance. The basic struture of the C.III remained similar to the C.I, being a conventional wood frame covered in fabric. However, the aerodynamic design, wing planform, and airfoil section were all changed to provide better high-altitude performance. Development of the new wing cellule paced the C.III design.

The C.III was powered by the 200/220 hp Benz Bz.IV engine, providing it with substantially more power than the C.I, again to improve speed, climb, and ceiling. In addition to the more powerful engine, the nose entry was more streamlined for lower drag, and included a large propeller spinner. The prototype featured air intakes under the nose, but these were omitted in the production aircraft.

Unlike the earlier C.I, which had a fixed fin, unbalanced control surfaces, and good handling qualities, the prototype C.III was built with all control surfaces aerodynamically balanced to reduce control forces. To minimize weight and drag for maximum performance, the prototype C.III omitted

Above: A Rumpler C.III prototype is seen here in front of the Rumpler factory. Both pilot and observer have headrests for improved streamlining. All controls were aerodynamically balanced.

a fixed vertical fin, something tried by many German designers. However, few designs without a fixed fin were successful due to inadequate directional stablity, and the C.III was no exception. During its production run the C.III was subject to continuing revision to its control surfaces to improve its flying qualities. Like the prototype, all control surfaces on early production C.III aircraft were aerodynamically-balanced, and there was no fixed fin. Mid-production C.III aircraft changed to a fixed fin and unbalanced rudder to improve flying qualities. The ailerons were revised at the same time; when C.III aircraft were fitted with a fixed fin the ailerons were changed to unbalanced. Initially the balanced elevators were retained. The final production C.III aircraft eliminated the elevator balances to further improve flying qualities. There are several photos of crashed early configuration C.III aircraft after control problems, indicating the C.III control changes were necessary.

The C.III was fitted with the standard C-type armament, a fixed, synchronized gun for the pilot and a flexible gun for the observer. To make the type more versatile, a small bomb bay was built into the starboard side of the observer's cockpit. The prototype C.III had an unusually shaped turtledeck behind the observer's cockpit; this may have obstructed the oberver's field of fire and was replaced in production aircraft by a conventional turtledeck.

Rumpler C.III Specifications

Engine:	200/220 hp Benz Bz.IV	
Wing:	Span	12.66 m
	Area	34.8 m^2
General:	Length	8.20 m
	Empty Weight	839 kg
	Crew	160 kg
	Fuel	105 kg
	Military Load	164 kg
	Useful Load	520 kg
	Loaded Weight	1,264 kg
Maximum Speed (at 2,000 m):		136 kmh
Climb:	2000m	16 min
Service Ceiling:		4000m
Range:		480 km

The Rumpler C.III first appeared at the front at the end of 1916, and by the end of February 1917 42 were at the front. Although the C.III had better performance than the C.I, its handling characteristics were less pleasant and the control modifications mentioned above had to be introduced on the production line. Within two months the number of C.III aircraft at the front dropped to 22 as the C.III was quickly replaced by the more powerful Rumpler C.IV, and by the Autumn of 1917 the C.III had virtually disappeared from the front. Unlike the viceless C.I, the C.III did not go on to an extensive career as a trainer.

Above: The Rumpler C.III prototype is ready for a flight with both crewmen. The streamlined nose entry, including large propeller spinner, shows the design team's search for better performance through reduced drag.

Production and Service

The Rumpler C.III was built in small numbers. In July 1916 150 aircraft were ordered, but only 49 were built with serials 1863/16 to 1911/16. Serial 1912/16 was the prototype C.IV. The other 100 aircraft from that order may have been changed to the C.IV. From photographic evidence the transition from the early configuration (no fixed fin, all controls balanced) to the final configuration (fixed fin, all control surfaces unbalanced) took place at some point between serials

Above: The Rumpler C.III prototype seen from the front quarter shows its marked resemblance to the later C.IV. Several cooling intake vents underneath the nose differentiate this prototype from the production C.III and later C.IV. Test instruments are mounted on the front, outboard, starboard interplane strut. The propeller was made by Axial.

1869/16 and 1898/16 (although 1903/16 retained the balanced elevators of the mid-production configuration; apparently serial numbers were not always assigned in production order).

Only one C.III was at the front at the end of 1916, but two months later the inventory had jumped to 42. Thereafter C.III numbers at the front went into rapid decline as the more powerful Rumpler C.IV, which possessed significantly better performance, replaced it. The Rumpler C.IV quickly gained a reputation as the best German long-range reconnaissance aircraft, a reputation the C.IV and its derivatives retained throughout the war. All the *Rubild* Rumplers used the C.IV airframe, which was essentially the final production C.III airframe with unbalanced control surfaces.

In Retrospect

Development of the Rumpler C.III, especially the wing cellule, to achieve the best high-altitude performance was challenging. Handling and structural problems plagued production C.III aircraft, requiring a series of revisions to its control surfaces during production. Moreover, a more powerful engine than the 200/220 hp Benz Bz.IV was needed to achieve the airframe's full potential.

Thus from an operational perspective, the Rumpler C.III was, at best, a marginal success.

However, from a development perspective, the C.III was a great success. It was the operational test-bed for the airframe that, once developed, lead to the legendary C.IV and its related family of aircraft.

The Rumpler C.III was thus a transitional design between the highly-regarded, easy to fly general-purpose C.I and the high-performance C.IV. Although the C.III served at the front in modest numbers, its most significant role was that it introduced and refined the basic airframe used by the more powerful C.IV and its family of derivatives, which were the best German long-range reconnaissance aircraft of the war.

Above: The supplementary air intakes seen beneath the nose on the Rumpler C.III prototype in its original configuration have now been eliminated, and only a modest grill beneath the nose remains. The turtle-deck is also conventional now; this appears to the configuration for the early production C.III.

Above: This side view of the Rumpler C.III prototype emphasizes its clean lines. Although all controls are aerodynamically balanced, the curved rudder without fixed fin is clearly too small. The photo was taken on 22 August 1916.

Below: Only the under-nose cooling air intakes and radiator over the engine mar the Rumpler C.III prototype's clean lines. Test instruments are mounted on the front, outboard, starboard interplane strut.

Rumpler C.III First Prototype

Rumpler C.III Second Prototype

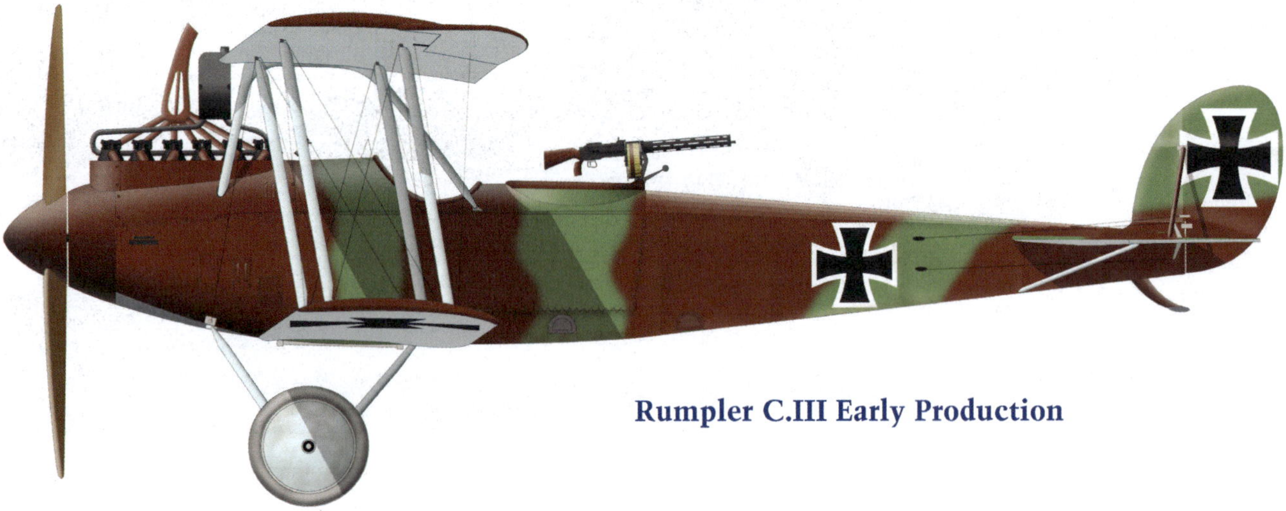

Rumpler C.III Early Production

Above: The unusual turtle-deck behind the observer's cockpit seen on the Rumpler C.III prototype in its original configuration has now been replaced by a conventional turtle-deck, most likely to give the observer a less obstructed field of fire. The aerodynamically-balanced ailerons and elevators are clearly visible.

Right: The Rumpler C.III prototype taking off on a test flight at the Rumpler factory.

Below: This appears to be an early production Rumpler C.III at the front. There is still no fixed fin but the turtle-deck is conventional and the supplementary cooling intakes beneath the nose seen on the prototype are also gone. The iron cross national insigna now have white outlines instead of being applied over a square white background. The ailerons and elevators retain their aerodynamic balances.

Above: Early production Rumpler C.III 'F15' is from the wireless operator training school. Again it has no fixed fin and no supplementary cooling intakes beneath the nose. The iron cross national insigna have white outlines and the overall color is darker than the typical Rumpler factory camouflage of the time. The dark characters 'F15' on the fuselage side have been outlined in white for greater visibility.

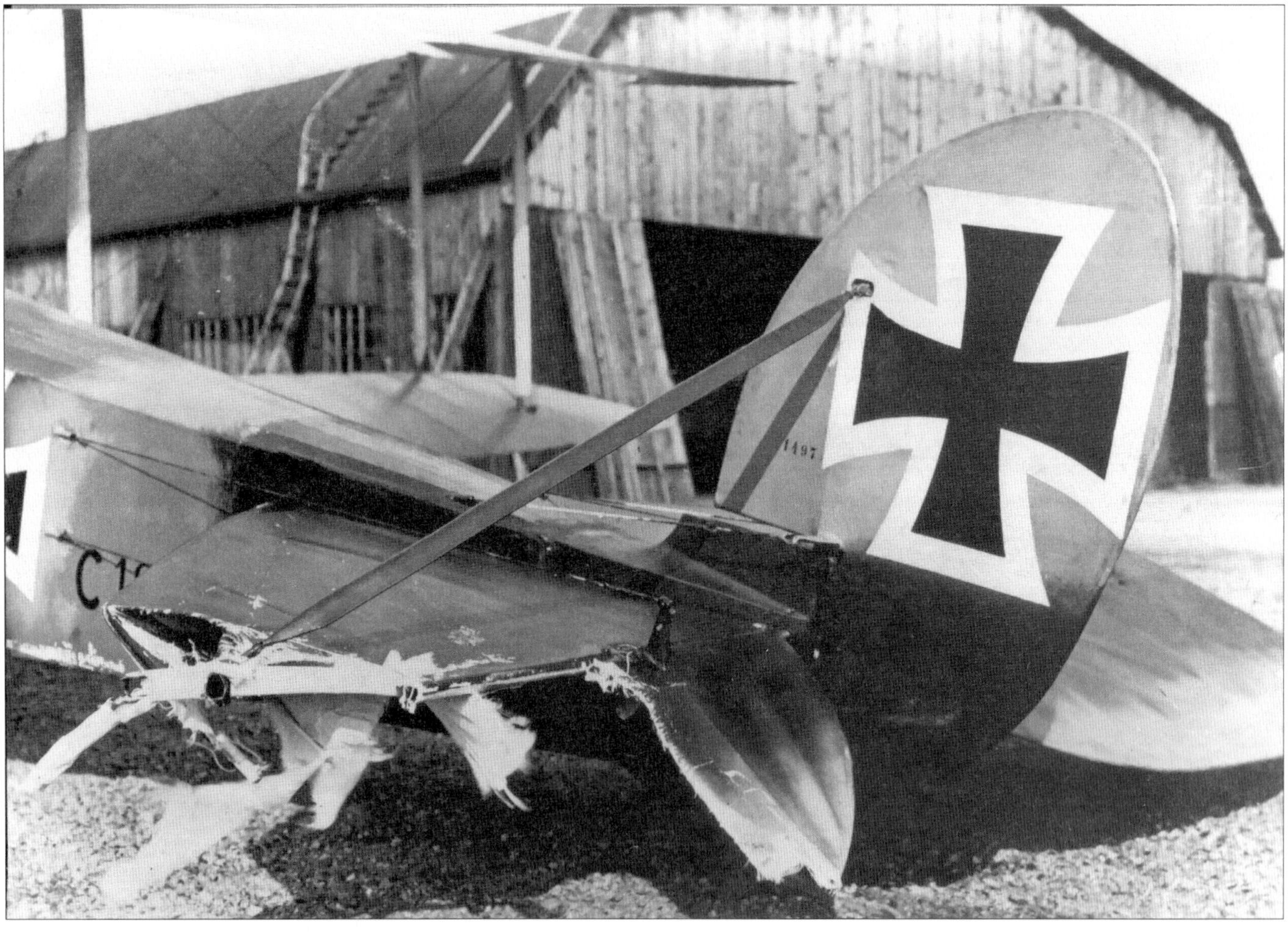

Above: *Feldwebel* Niemeier of *Fl.Abt. (A)* 250 suffered control failure during a flight in this early-production C.III, w/n 1497. Unfortunately, the military serial number is not visible in the photo. The C.III was designed as lightly as possible for the best climb and ceiling, and numerous modifications had to be made to strengthen the airframe based on experience.

Above & Below: On 17 January 1917 pilot August Quoos suffered control failure during a test flight near Colmar in this production C.III, work number 1466, and suffered this crash landing as a result. Unfortunately, the military serial number is not visible in either photo. See *Cross & Cockade* Volume 7, No.2, 1966, for Quoos' story of this incident.

Above & Below: Early-production C.III work number 1474 has suffered a bad landing – and again the military serial number is not visible in the photos! The large number of crash photos of production Rumpler C.III aircraft despite the relatively small number built is an indication that further development of the control system and flying qualities was needed.

Above & Below: Early-production Rumpler C.III 1869/16, work number 1470, is down in the snow, likely as a result of a bad landing. The pilot's fixed gun, factory camouflage and aerodynamically-balanced controls are clearly visible.

Above: An early-production C.III has suffered a rough landing. This view shows a window in the cockpit floor and a control rod under the fuselage.

Below: Late-production Rumpler C.III 1898/16 after it landed near Elverdinghe on 31 August 1917.

Above & Below: These photos show Rumpler C.III 1898/16 after it landed near Elverdinghe on 31 August 1917 and was given the British captured aircraft designation G66. Examination of factory component markings revealed the upper wing was manufactured on 16 Nov. 1916 and the lower wing was manufactured on 27 Nov. 1916. Despite being slightly damaged, the camouflage and markings are nicely visible. This Rumpler C.III represents the final production configuration; all control surfaces are unbalanced. It looked very much like a Rumpler C.IV; the real differences were the engine and the squared-off lower wing tip planform so clearly shown below.

Above: Rumpler C.III 18xx/16 illustrates the intermediate production configuration. It has the revised vertical tail with fixed fin to improve stability, and both the ailerons and rudder are now unbalanced. However, the elevators are still balanced. This transition took place sometime between serial 1869/16, an early configuration machine, and 1898/16, representing the final production configuration.

Facing Page, Above & Below: Rumpler C.III 1903/16 shows its revised vertical tail with fixed fin and balanced elevators, representing the middle production configuration. Lightly built to achieve high altitude, it has suffered damage to the rear fuselage and balanced elevators. Both C.III and early C.IV aircraft, which shared their airframe, had to be modified at the factory to stengthen the rear fuselage. Later C.IV airframes included these changes in production.

Below: This late production Rumpler C.III with unbalanced elevators carries the unit insignia of *Flieger Abteilung* 19. Other than its use of the Benz engine, from the side it looks like a Rumpler C.IV. The lower wing tip shape, difficult to see clearly from this angle, is the most visible difference between the late C.III and early C.IV.

Above: This rare late production Rumpler C.III lasted at the front until early 1918 as indicated by the Balkenkruez insignia. The rudder is now white. It also carries the unit insignia of *Flieger Abteilung* 19. Is it the same aircraft shown on page 66 after re-covering and re-painting? The names Tetzner and Schneider were written on the back of the photograph; were these the crewmen? It is almost certainly the aircraft also shown below.

Below: Rumpler C.III 1873/16 of *Flieger-Abteilung* 19 rests between missions. Only one C.III remained at the front into 1918 and, from the national insignia, it was probably this aircraft. This may be the same aircraft shown on page 66.

Above: LVG C.II 2135/15 is at left foreground and three late-production Rumpler C.III aircraft rest in the right foreground on 16 April 1917. Interestingly, there are no insignia on the rudders of the C.III aircraft. A Rumpler C.I is in the center background with another LVG C.II at right background.

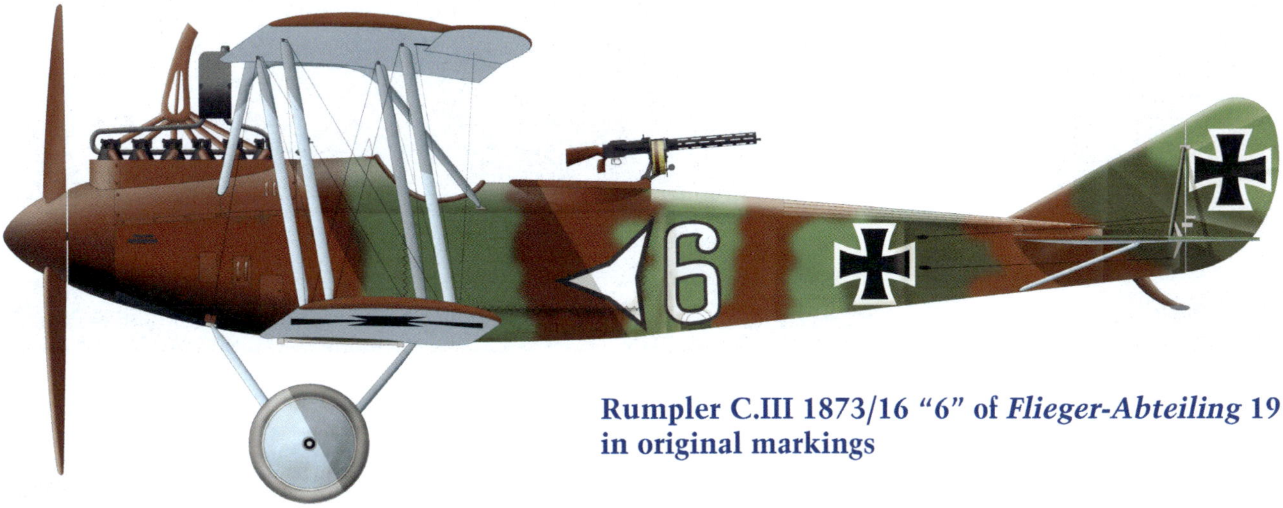

Rumpler C.III 1873/16 "6" of *Flieger-Abteiling* 19 in original markings

Rumpler C.III 1873/16 "6" of *Flieger-Abteiling* 19 in 1918 markings and, apparently, one-color upper surface camouflage

The camouflage patterns seen on the C.III and C.IV profiles are, in many cases representative, as the actual patterns could not always be discerned and in these cases are based on typical Rumpler practice as noted on similar aircraft.

Rumpler C.IV Family

Above: Early production Rumpler C.IV 6692/16 in factory finish. Other than its use of the Mercedes D.IVa engine and the rounded planform of the lower wings, from the side it looks very similar to a late-production Rumpler C.III. (The Peter M. Bowers Collection/The Museum of Flight)

Starting with its *Taube* series and continuing through its excellent C.I, Rumpler had made a name for itself. Production of the transitional C.III had been a challenging episode as Rumpler worked to sort out the airframe to achieve satisfactory airframe strength and flying qualities. Ultimately, the C.III was limited by its engine and more power was needed to get the full potential for its airframe. Availability of the new 260 hp Mercedes D.IVa was the solution. Rumpler combined that powerful new engine with the C.III airframe to create the milestone Rumpler C.IV, the airplane that made Rumpler famous during the war. The new engine, coupled with a refined airframe based on the late-production C.III, endowed the C.IV with unexcelled high-altitude performance and created a family of aircraft that were the best strategic reconnaissance airplanes fielded by Germany during the war.

The prototype C.IV, Rumpler type 6A7, which made its first flight in late September 1916 appears to have been closely related to the prototype C.III airframe, but production C.IV aircraft used the late-production C.III airframe with greater strength and better flying qualities. Construction was the typical wood, wire, and fabric for the time but carefully engineered for minimum weight. The most visible difference between the early production C.IV and late production C.III was the planform of the lower wing tips; that of the C.III was squared off, whereas the C.IV used a rounded 'dragonfly' wing profile. The rounded wing planform reduced the induced drag (drag 'induced' by lift due to creation of wake turbulence) and, together with the higher power of the new engine, gave greatly improved high-altitude performance.

On October 18, 1916, the C.IV prototype climbed to 5,000 m in 33 minutes with full combat load, including fuel for 3½ hours, and achieved a speed at that altitude of 165 km/h. This was very impressive performance for the time and the C.IV was quickly ordered into production. Type testing was done in mid-January 1917. On February 10 the C.IV was approved for production and service with the caveat that the heavy aileron control forces would be alleviated by the 101st production aircraft.

In late February the first Rumpler C.IV aircraft

Rumpler Aircraft Using the C.IV Airframe		
Designation	**Engine**	**Notes**
Rumpler C.IV	260 hp Mercedes D.IVa 245 hp Maybach Mb.IVa 260 hp Basse & Selve BuS.IVa	Developed from earlier C.III, many different equipment installations. Early production aircraft had propeller spinners.
Rumpler C.IV(Pfal)	260 hp Mercedes D.IVa	Originally designated Pfalz C.I, license-built C.IV with ailerons on all four wings for training use
Rumpler C.V	245 hp Maybach Mb.IVa	Bomb racks
Rumpler C.VI	260 hp Mercedes D.IVa	Equipped with Messter strip camera; later re-designated *Rubild*
Rumpler C.VII	245 hp Maybach Mb.IVa	
Rumpler C.VIII	180 hp Argus As.III	Operational trainer not intended for combat. Different lower wings
Rumpler C.IX	245 hp Maybach Mb.IVa	Long-range fuel tank
Rubild	260 hp Mercedes D.IVa	New designation for C.VI
Rubild Mb	245 hp Maybach Mb.IVa	Rubild equipped with Maybach engine
The Rumpler C.IV was modified with a number of different engines and carried a wide variety of equipment. Inititally these aircraft were all designated Rumpler C.IV, but in late 1917 new designations were applied to aircraft fitted with different varieties of engines and equipment. This table summarizes these variations of engines, equipment, and associated designations for the aircraft that were all based on the same airframe.		

Above: Side view of an early Rumpler C.IV prototype, perhaps the first, shows the unusual rear fuselage fairing behind the observer's cockpit that was later removed. There is no propeller spinner and no opening for a fixed gun for the pilot. It looks very similar to the Rumpler C.III prototype and was clearly designed at nearly the same time.

Rumpler C.IV Specifications

Engine:	260 hp Mercedes D.IVa	
	260 hp Maybach Mb.IVa	
Wing:	Span (Upper)	12.66 m
	Span (Lower)	12.1 m
	Chord (Upper)	1.83 m
	Chord (Lower)	1.33 m
	Gap	1.85 m
	Area	34.8 m²
General:	Length	8.40 m
	Height	3.20 m
	Empty Weight	1,017 kg
	Useful Load	580 kg
	Loaded Weight	1,630 kg
Maximum Speed:		175 kmh
Climb (Mercedes):	1000m	4 min
	2000m	7.5 min
	3000m	13 min
	4000m	19 min
Climb (Maybach):	1000m	3 min
	2000m	6.5 min
	3000m	11.5 min
	4000m	16.5 min
	5000m	24.5 min
	6000m	37.5 min
Service Ceiling:		6800m
Range:		585 km

Rumpler C.IV Family Orders

Date	Qty	Serials	Notes
	1	1912/16	C.IV Prototype
Sep. 1916	100	8203–8301/16	C.IV
Sep. 1916	150	6675–6824/16	C.IV
Oct. 1916	150	7500–7649/16	Pfalz-built C.IV
Nov. 1916	150	8403–8558/16	C.IV
Nov. 1916	50	8900–8949/16	Bayru-built C.IV
Jan. 1917	150	See Note 1	Pfalz-built C.IV
Feb. 1917	100	1420–1519/17	C.IV
Feb. 1917	100	See Note 2	Bayru-built C.IV
June 1917	100	6400–6499/17	C.VII
July 1917	100	7600–7699/17	C.IV & C.VII
July 1917	400	7950–8349/17	C.IV/C.VI/C.VIII
July 1917	200	Unknown	Bayru-built C.IV
May 1918	200	3150–3349/18	See Note 3
July 1918	100	4000–4099/18	See Note 3
July 1918	100	7900–7999/18	See Note 3
Oct. 1918	50	12200–12249/18	Rumpler C Mb

Notes: 1. Serials 550–699/17 not confirmed.
2. Known serials 2746–2820/17.
3. Combined C.IV, C.VII, Rubild, Rubild Mb

reached the front. *Flieger-Abteilung* 18, for example, received five Rumpler C.IV aircraft as early as February 21. *FA* 18 and other units equipped with the C.IV quickly acknowledged the excellent high-altitude and long-range performance of the C.IV. However, they also complained that the lightly-built airframe was not robust enough to endure the strong engine vibrations, prevent aerodynamic forces from twisting the fuselage during maneuvering, and preclude rear fuselage factures from the rigors of front-line use, all issues that first appeared in the Rumpler C.III. Despite testing that showed the fuselages to be sufficiently strong according to the specification then used, both the C.III and C.IV were withdrawn from the front in mid-April 1917 and returned to the factory for re-building. The rear fuselages were reinforced and plywood panels replaced much of the fabric.

Aerodynamic research continued as aircraft were produced, and wind tunnel testing revealed that lower drag and higher speed could be achieved by eliminating the propeller spinner. This counter-intuitive result was applied to the production line starting in May 1917, and later production C.IV aircraft – and all the derivative types listed in the family table – were built with a rounded nose without spinner. The result was a welcome 10–15 km/h speed increase.

Early production C.IV aircraft were equipped with a control stick for the pilot. For some reason this was not fully satisfactory, and beginning in May–June 1917 production aircraft had a control wheel.

In action the Rumpler C.IV promptly made a name for itself as the C-type with the best performance. Together with the robust, tractable DFW C.V, the Rumpler was the most useful German two-seater due to its high performance and good maneuverability. However, the Rumpler C.IV also proved itself to be a handful for pilots due to its slow aileron response and propensity to stall, which was often abrupt and unforgiving. One Rumpler pilot ferrying his aircraft to the front spun from 3,200 m to 800 m before he could recover, a very large altitude loss in a biplane.

The C.IV was modified by the substitution of different engines to further improve ceiling and speed at altitude and also used different cameras, equipment, and long-range fuel tanks for enhanced strategic reconnaissance capability. These modifications included use of the 245 hp Maybach Mb.IVa over-compressed engine starting in April 1917. The ceiling of C.IV aircraft with this engine was eventually raised to more than 7,000 m. In late 1917 the 260 hp Básse & Selve BuS.IVa became

Above: Early production Rumpler C.IV 6745/16 in factory finish shows its clean lines.

Below: Early production Rumpler C.IV 6820/16 in factory finish prepares for another mission.

Above: Later production Rumpler C.IV 8476/16 in factory finish has the new nose contours without spinner. Eliminating the spinner reduced drag and added 10–15 km/h to the aircraft's speed.

available and *Idflieg* allocated these engines to the Bayerische Rumpler Werke (Bayru); these aircraft were sometimes known as the Rumpler C.IV(BuS). Bayru also built the C.IV with the original Mercedes engine. Interestingly, the 260 hp Mercedes D.IVa, the first engine installed in the C.IV, had to be rotated clockwise (as seen from the cockpit) a few degrees to fit without its intake manifold interfering with the left cabane struts.

At first there were no designation changes to reflect these modifications, but the designations of later production aircraft were changed for this reason. The table on page 71 summarizes the various designations and modifications of aircraft in the C.IV family that used the C.IV airframe. The Pfalz-built Rumpler C.IV(Pfal) was the only type that could be easily distinguished from the basic C.IV because it had ailerons on all four wings to improve roll control. The ailerons were connected by actuating struts. The C.IV(Pfal) was intended as a trainer – the reason its aileron control was improved – but a few apparently saw operational service at the front.

In fact, unless a photograph shows the designation painted on the aircraft, these aircraft (other than the C.IV(Pfal)) are virtually indistinguishable from the basic C.IV. The only visible difference between the various types is the visible portion of the engine, which is difficult to distinguish with certainty. This is a major contributor to the fact that the later members of the C.IV family are not well known. Another contributor is that they were seldom met in combat because Allied fighters only rarely could intercept them. Even when an interception occurred, they looked like the normal C.IV to the Allied pilot. A final contributor to the identification challenge is

Above: Early production Rumpler C.IV 6770/16 and crew.

Above: Later production Rumpler C.IV 1449/17 has suffered a landing accident. It wears standard factory camouflage enhanced with red/ white/black (German national colors) markings on the fuselage, wheel covers, wing struts, and nose, making it much more colorful than most Rumplers; see the below color profile.

Rumpler C.IV C.1449/17 of an unknown unit was a very colorful aircraft.

that different types were built together within the same production batch and range of serial numbers. Engine installation depended on availability, and the airframe was modified on the production line as necessary for the engine and equipment to be installed. If airframes and serial numbers were documented according to aircraft type, which depended on the engine and equipment fitted, this information has not come to light.

The first three C.IV aircraft equipped with the Messter strip cameras were so modified in April 1917, but factory deliveries of *Reihenbildflugzeuge* (strip photography aircraft) did not begin until that October. These aircraft flew at altitudes up to 7,000 m where the air temperature was -50° Centigrade on flights 5–6 hours long. The extreme cold experienced by the aircrews required electrically-heated flying suits and primitive oxygen breathing apparatus.

In July 1917 aircrews started to report being attacked by single Allied fighters at these high altitudes; some responses included removing the pilot's fixed gun to reduce weight and polishing the wing fabric to increase speed. The Allied fighters involved were certainly powered by the Hispano-Suiza V-8 as rotary-powered fighters did not have the ceiling to reach a Rumpler.

Top British ace James McCudden specialized in intercepting Rumplers and stated that his personal attention to his SE.5a (he was an aircraft mechanic before becoming a pilot) enabled him to fly much higher than the typical SE.5a, which could rarely climb above 17,000 feet in winter; he was thus able to intercept surprised Rumplers that the typical SE.5a could never reach. Once attacked, experienced Rumpler aircrews often saved themselves by retaining their high altitude, at which they were as fast as an SE.5a. If the Rumpler pilot lost altitude during the combat the fighter gained the advantage and had a much greater likelihood of downing the Rumpler.

In Retrospect

The Rumpler C.IV and its derivatives based on its airframe were the best German high-altitude reconnaissance airplanes and are among the most famous aircraft of the war. Edmund Rumpler and his company are famous above all for the performance of the exceptional C.IV family.

Above & Right: Rumpler C.IV 1454/17 of *FA(A)* 209 was brought down by *Sous-Lt.* Covin of *SPA* 31 on Oct. 24, 1917; its crew was made PoW. Here it is seen after French markings were applied. A small, non-standard pointed spinner is fitted to the propeller; the unit may have added this.

Right: Rumpler Rubild 8322/17 in factory camouflage and wearing the straight-sided 1918 national insignia.

Below: Rumpler C.IV 7610/17 in American hands in 1918 as shown by the straight-sided *Balkenkreuz*, apparently after repairs. It had the Mercedes D.IVa engine typical of the C.IV.

Below: Both ailerons of Rumpler C.VII 8107/17 are drooping, indicating the control cables are disconnected. Late-style national insignia date the photo in 1918.

This Page & Facing Page: Rumpler Rubild Mb 8186/17 flown by pilot *Lt.* Oskar Seitz. Seitz was instumental in designing the custom observer's gun mounting that combined a captured Lewis machine gun with the standard Parabellum for additional firepower to defend against Allied fighters. The aircraft has lozenge camouflage fabric on the wings and horizontal tail with painted fuselage and late-style national insignia. The "1" on the fin was a tactical marking.

1
8186

Rumpler Rubild Mb 8186/17 flown by *Lt.* Oskar Seitz with dual guns for observer

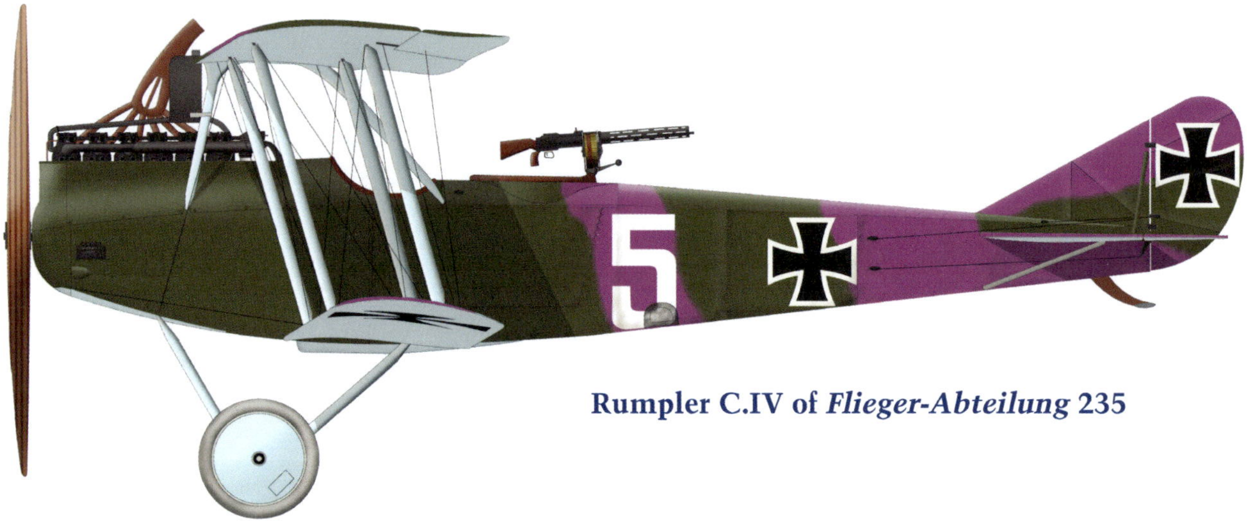

Rumpler C.IV of *Flieger-Abteilung* 235

Rumpler C.VI C.8254/17

Above: Another view of Rumpler Rubild Mb 8186/17 flown by pilot *Lt.* Oskar Seitz.

Below & Below Right: More details of the custom double gun mount fitted to Rumpler Rubild Mb 8186/17.

Below: Rumpler C.IV 7955/18 serving with *FA*304b after a bad landing.

This Page & Facing Page: Rumplers flew long missions at extreme altitudes, and their crews were subjected to sub-zero cold in open cockpits, so heavily-insulated flying gear, often electrically heated, together with protective masks were needed to prevent frostbite. These scenes of crewmen preparing for flight illustrate some of their specialized flight equipment, including electrically-heated flight suits, gloves, and face masks.

Above: Rumpler C.VI 8254/17 has a flare rack for the observer and is coverd with printed camouflage fabric.

Above: This Rumpler C.IV crew is getting a final briefing before their mission. The observer must be a good marksman because his gun has a telescopic sight for defending the aircraft at long range.

Above: Rumpler C.IV 8476/16 in its factory painted camouflage.
Below: Another Rumpler observer whose gun has a telescopic sight.
Bottom: Control wheel in a Rumpler cockpit; early-production aircraft had a control stick for the pilot.
Below Right: Confident flight crew photographed by their Rumpler.

Above: The crew of this Rumpler C.IV appeal for kindness with their markings "Good People don't shoot". In case that appeal does not work the observer has a machine gun on an unusual straight mounting. One pilot told the author he chose to fly Rumplers instead of fighters because he "just didn't have the heart to dive down on an unsuspecting enemy and shoot him down."

Facing Page, Above: The crew of this Rumpler C.IV delivers Christmas dinner. This photo gives a better view of the unusual mounting for the observer's flexible gun.

Facing Page, Below: Another view of a Rumpler C.IV delivering packages to the front. Despite use of the unusual observer's gun mounting, which must have caused more drag than standard mountings, this appears to be a different aircraft, or at least a different occasion, than that above.

Rumpler C.IV
C.8288/16

Right: Rumpler C.IV C.8288/16 wears a skull and crossbones marking, something seen far more commonly on fighters than reconnaissance airplanes. The proud crew is enjoying the moment.

Above, Above Right, Below: The unusual observer's gun mounting is visible on this crashed Rumpler, supporting its identification as an early production C.IV.

Right: Close-up view of a Rumpler C.IV; the unusual straight-tube observer's gun mount is visible.

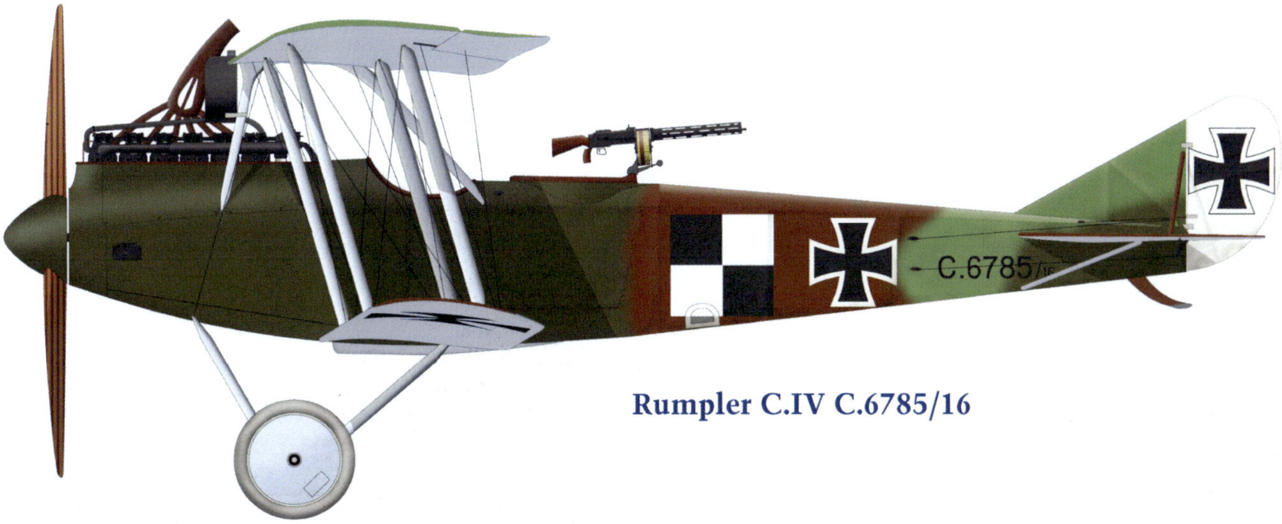

Rumpler C.IV C.6785/16

Rumpler C.IV C.6715/16 of *Flieger Abteilung* 19

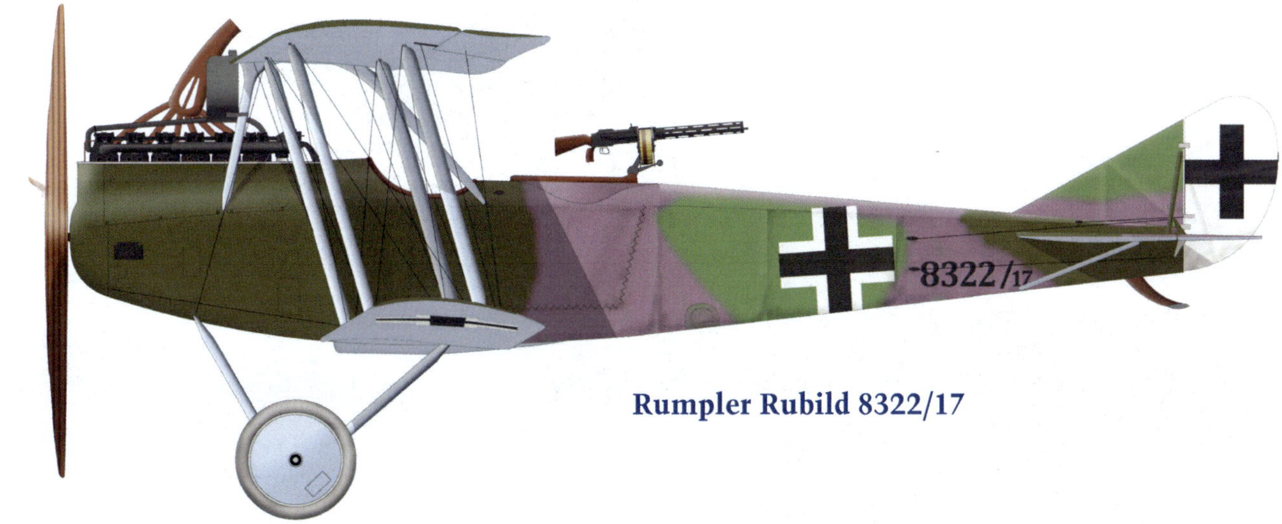

Rumpler Rubild 8322/17

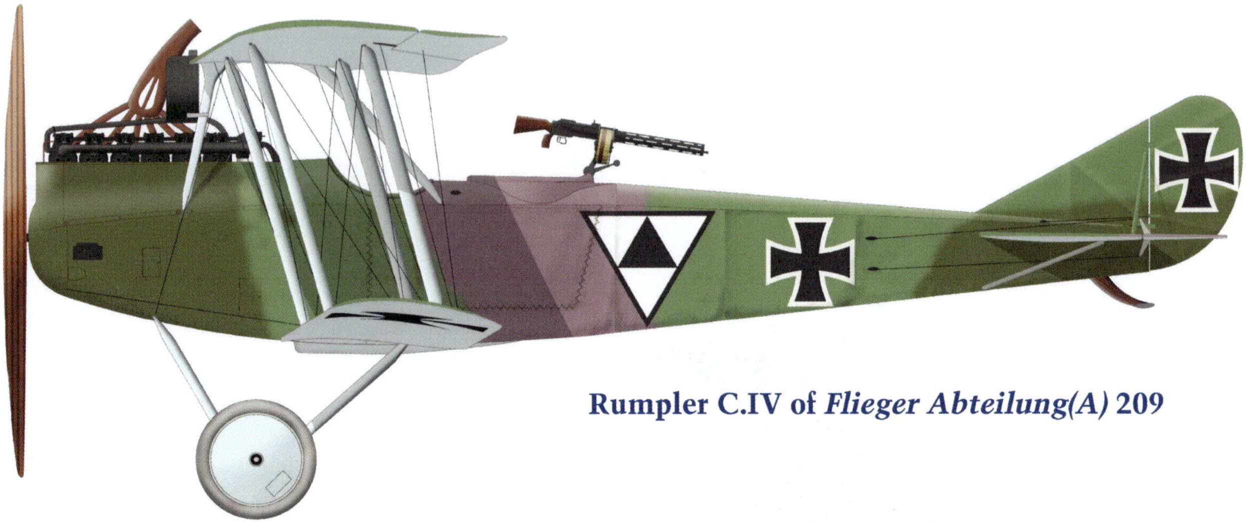

Rumpler C.IV of *Flieger Abteilung(A)* 209

Rumpler C.IV of *Flieger Abteilung(A)* 209

Rumpler C.IV in camouflage printed fabric

Above: Rumpler C.IV of *Flieger-Abteilung* 18; *Lt.* Heinz Lubarsch was the observer.

Right: Rumpler C.IV of *Flieger-Abteilung (A)* 209 carries a wicker supply cannister. When dropped the canister descended by parachute. *Flieger-Abteilung (A)* 209 also used AEG N.I night bombers for this purpose.

Below: Rumpler C.IV on a snowy airfield at Minsk, Russia.

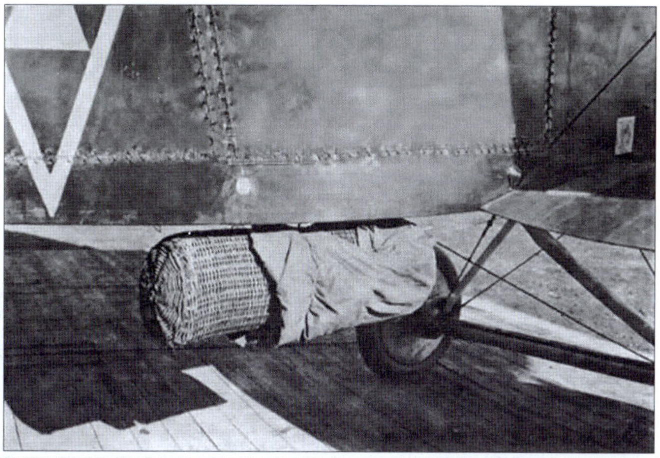

Left: Rumpler C.IV 8485/17 of *FA(A)* 219 wears an interesting camouflage scheme. Like the earlier C.III from which the C.IV got its airframe, the C.IV was lightly designed for minimum weight to maximize climb and ceiling. The result was the airframe had to be locally strengthened a number of times based on operatonal experience.

This Page, Lower Photos on Facing Page: Is this a case of before and after? Rumpler C.IV tactical number '3' of *Flieger-Abteilung* 7 with crew on facing page above and after coming down behind Allied lines in other photos, although the first photo may be a different aircraft. Rumpler C.IV C.8500/16 with 260 hp Mercedes D.IVa was captured on 29 December 1917, brought down at Monchy by ground fire. Given the capture number G.117, it was closely examined and was the basis of the report in *Jane's All the World's Aircraft*. It is in the early factory finish plus the unit insignia of the arrows on the fuselage sides and top. The eyes painted on the nose are thought to be individual markings. Unfortunately, the photos were taken after the iron cross national insignia had been painted over with roundels.

Above: Rumpler C.IV tactical number '4' of *Flieger-Abteilung* 7 after a rough landing.

Below: Maybach-engined Rumpler C.IV 1423/16 tactical number '4' of *Flieger-Abteilung* 7. Although having the same tactical number, this may be a different aircraft then that shown above; notice the arrow markings on the upper wings not on the airplane above. Of course, the upper wing arrows may have been added after repair. From this batch onward the type designation, in this case 'C.4', was added to the serial number to distinguish the sub-types of the family. Roman numerals were used after this batch.

Above: Early-production Rumpler C.IV in factory finish.

Below: A Rumpler C.VII with Swastika personal marking and massive camera between the cockpits.

Below Right: Crash of Rumpler C.IV 8512/16. The Swastika is aft of the one on the C.VII.

Below: Printed lozenge fabric on a late-production Rumpler.

Left: This early-production Rumpler C.IV retains the straight bar observer's gun mount.

Above: *Lt.d.R.* Karl Kerner (at right) and *Uffz.* Prantzsch (at left with laurel wreath) of *FA(A)* 282 stand in front of their Rumpler, perhaps a Rubild, after their 100th mission. Kerner wears an electrically-heated flying suit.

Left: This late-production Rumpler had a wind-driven electrical generator to power the crew's electrically-heated flying suits.

Above: Two Rumplers are lined up in this well-known photograph with a Hannover CL.II and DFW C.V in the background. Unusually, the early Rumpler nearest the camera, C.IV 8267/16, is covered with printed camouflage fabric. The second Rumpler, C.IV 6695/16, was sent back to "Lager West" from *AFP* 6 in April 1917 for strengthening of the rear fuselage.

Left: After the war Rumpler used both C.I and C.IV aircraft in a short-lived airline, *Rumpler Luftverkehr* (Rumpler Air Traffic). Here D72 shows its heritage of clean lines in March 1919. Its route of Berlin-Augsburg-Munich is painted on the fuselage.

Above: Early-production Rumpler C.IV with flight crew and early national insignia.

Right: Early-production Rumpler C.IV after a bad landing. The aircraft has the typical factory painted camouflage finish for early production aircraft.

This Page & Facing Page: Late production Rumpler Rubild Mb 12212/18 in British hands after the war. It has the late factory finish of overall lozenge printed fabric.

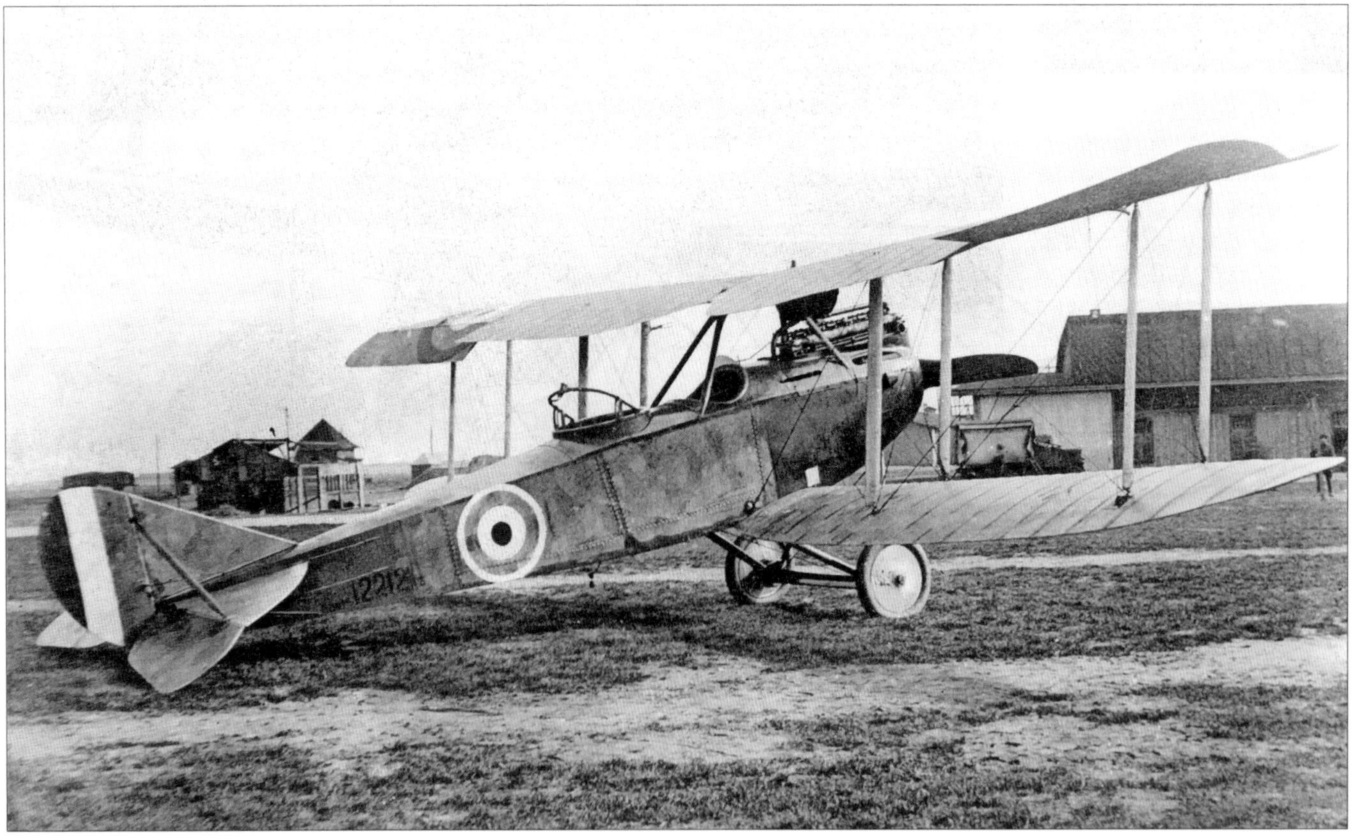

This Page & Facing Page: Late production Rumpler Rubild Mb 12212/18 in British hands after the war. Doubtless this was a welcome opportunity for some British fighter pilots to finally get a close look at an airplane type few of them had ever seen at close range due to its exceptional ceiling and speed at altitude.

Above: Late production Rumpler Rubild Mb 7955/18 delivered to Czeckoslovakia after the war.

Left: Rumpler Rubild Mb 8231/17 in French hands postwar.

Left: Rumpler C.IV x693/1x, probably 6693/16, of *FA(A)* 209, brought down behind French lines. When Rumpler C.IV aircraft were used for general two-seater duties at lower altitudes they were more vulnerable to fighters and anti-aircraft fire.

Rumpler C.IV C.8500/16 of *Flieger-Abteilung* 7

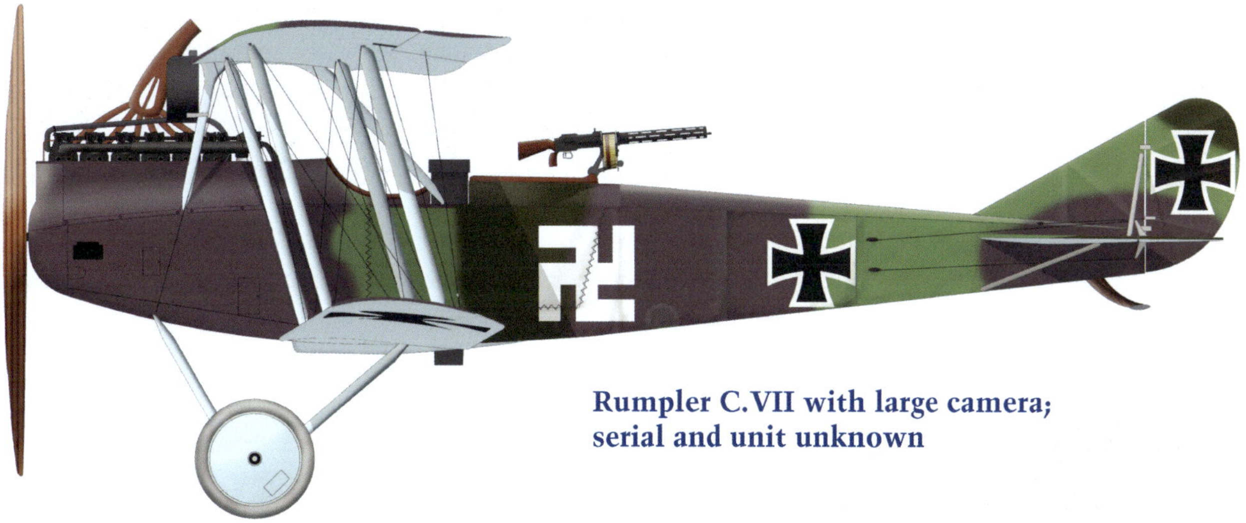

Rumpler C.VII with large camera;
serial and unit unknown

Rumpler Rubild Mb 8222/17

Above: Rumpler Rubild Mb 8222/17 landed in Switzerland, perhaps in preference to delivering the aircraft to the Allies.

Above: Colorfully-marked Rumpler in late insignia.

Below & Right: Late production Rumpler C.VII 7927/18 in British hands after the Armistice.

Right: Rumpler C.IV 8424/16 *Dalila* of the first production batch undergoes maintenance. This early production C.IV is the subject of Steve Anderson's cover painting. LVG C.V aircraft are on both sides of *Dalila*.

Above: Both reconniassance units and fighter units had unit markings, but personal insignia were more common in fighter units. Here a Rumpler in early factory camouflage, with paint chipping off the fabric, proudly displays a lion personal marking. Details of the unit and crew are not known, but one wonders if this was a Bavarian lion.

Left: Rumpler C.IV with early national insignia and painted camouflage is lined up with the rest of its unit.

Below: Crudely retouched postcard of Rumpler C.IV 8428/16 (w/n 2276) captured by the French. The aircraft was on display at the Place Jean Bart in Dunkerque. As evidenced by the white arrow unit marking this was a *Fl. Abt.* 7 aircraft. It was thoroughly studied by the French.

Right: Late-production Rumpler C.VII covered overall with printed camouflage fabric.

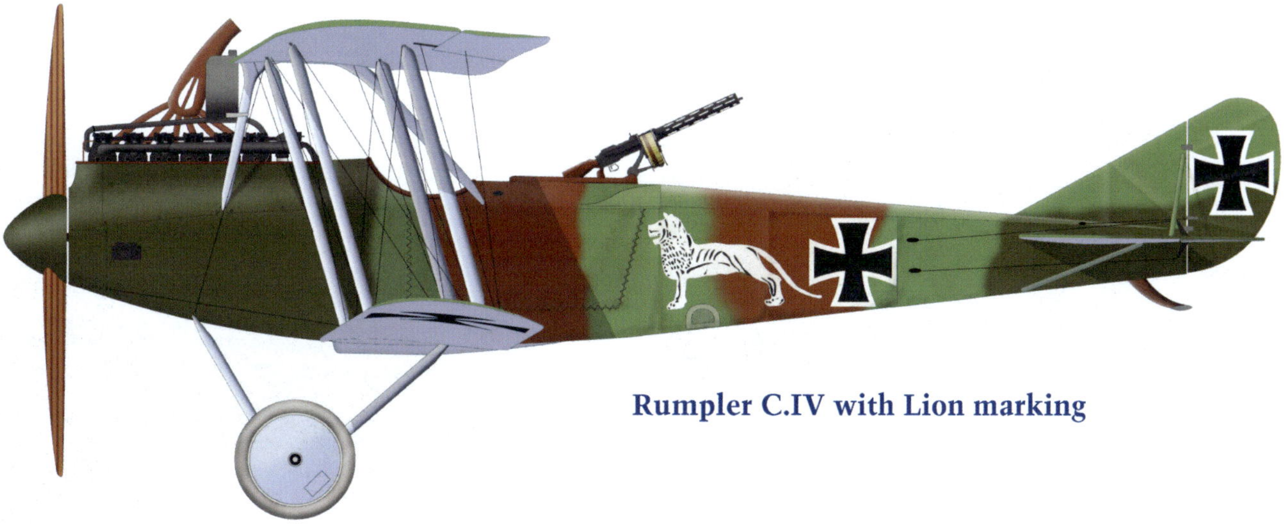

Rumpler C.IV with Lion marking

Rumpler C.IV C.8461/16 of *Flieger-Abteilung* 208

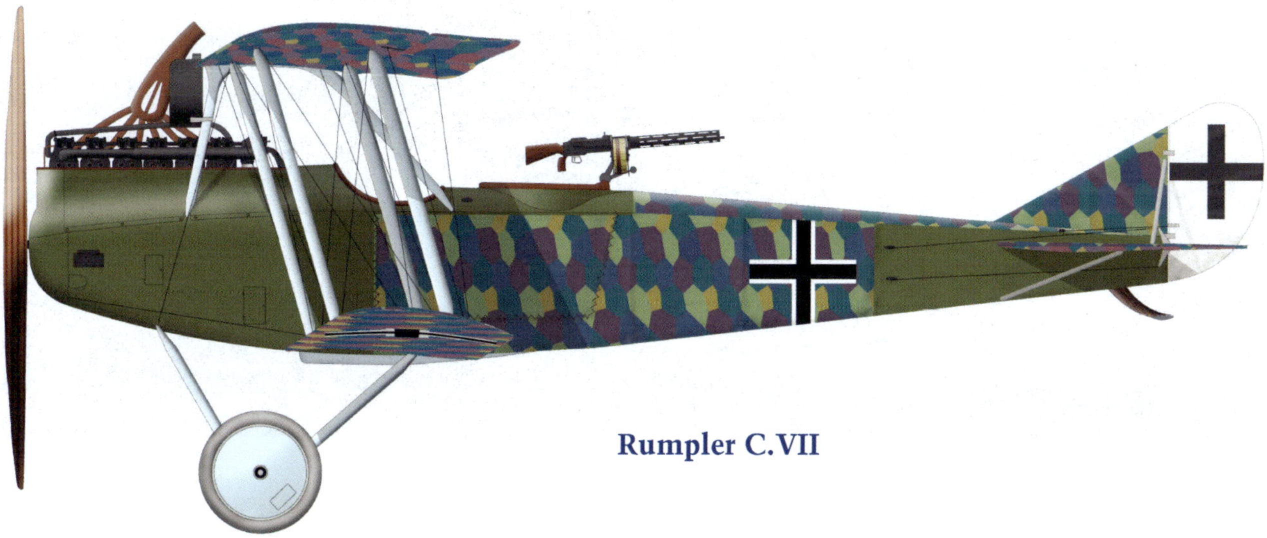

Rumpler C.VII

Above & Below: Late-production Rumpler C.VII covered with printed camouflage fabric.

Below: This early-production Rumpler C.VII in Allied hands wears painted camouflage.

Above: A nonchalant crew photographed with their Rumpler. The arrow marking on the side and top of the fuselage is the unit marking of *Flieger-Abteilung* 7.

Below: The ground crew photographed with their Rumpler. The triangle marking on the fuselage side appears to be the unit marking of *Flieger-Abteilung (A)* 209, although the colors are reversed from their usual presentation.

Above: An early-production Rumpler C.IV is a fine backdrop for this group portrait; another rests in the background.

Below: A mid-production (no spinner but wearing painted camouflage instead of fabric printed with the lozenge pattern) Rumpler seems to be the center of attention for this busy gathering. Crewmen in flight clothing indicate it is either ready for flight or has just returned.

Above & Below: An experimental Rumpler C-type with a massive, 345 hp Austro-Daimler V-12 engine installed. Fitted with this powerful engine provided by Austria-Hungary, the airplane was delivered to the Austro-Hungarian *Luftfahrtruppen* and given serial number 00.01. It crashed and was not repaired. Unfortunately, there is no record of its performance with this powerful engine, but it may have been somewhat nose heavy.

Rumpler C.VIII

The Rumpler C.VIII was a training version of the Rumpler C.IV. To conserve high-power engines it used a180 hp Argus As.III instead of the 260 hp Mercedes. The lower wing profile was similar to the Rumpler C.I and C.III rather than being rounded like the C.IV. The C.VIII was intended to give trainees the experience of handling a high-performance two-seater before they reached operational units instead of relying on obsolete aircraft. The C.VIII passed its *Typenprüfung* (type test) on 24 November 1917, some 200 being ordered initially (12400/17 to 12599/17). Postwar the C.VIII was used by other countries, especially the Netherlands, which purchased 40 aircraft in April 1918.

Above & Below: The Rumpler C.VIII was a trainer derived from the C.IV but with a lower-power engine not in high demand at the front. It was designed to give trainee pilots the experience of handling a high-performance two-seater before arriving at their front-line unit. The lower wingtip was squared off, not rounded like that of the C.IV.

Rumpler C.VIII Specifications		
Engine:	180 hp Argus As.III	
Wing:	Span (Upper)	12.22 m
	Span (Lower)	10.40 m
	Area	35.7 m²
General:	Length	7.10 m
	Height	3.20 m
	Empty Weight	874 kg
	Useful Load	580 kg
	Loaded Weight	1,374 kg
Maximum Speed:		145 kmh

This Page: Rumpler C.VIII trainers serving with the Netherlands. The Netherlands was a major user of the Rumpler C.VIII and purchased 40 aircraft before the Armistice. Germany sold these aircraft to purchase 5,000 horses from the Netherlands.

Rumpler 7C1

Above: The first version of the Rumpler 7C1 without fixed fin is seen in the Rumpler factory. The Rumpler 7C1 and 7D1/2 prototypes were a search for the ideal aerodynamic form.

"Everything is set on achieving the greatest possible performance, particularly with regard to climbing ability and speed." (Edmund Rumpler)

Extreme designs and their performance indicate the leading edge of technology for their time. However, racing automobiles and record-breaking aircraft are rarely practical on a daily basis; the main production emphasis is generally on what is practical and useful. In the case of long-range reconnaissance aircraft that must have the speed, climb, and ceiling to avoid interception, the leading practical WWI German aircraft was the Rumpler C.IV and its derivatives that shared its airframe. Evolution of the C.IV airframe into something lighter, more compact, and streamlined culminated in the excellent Rumpler C.X that appeared too late to see operational service. This section profiles another Rumpler design for the same role that pushed the state of the art too far. Like its single-seat counterpart the 7D1, the Rumpler 7C1 could not be developed into a practical production aircraft and is an example of what is sometimes ruefully known as the 'bleeding edge' of technology.

Postwar Edmund Rumpler stated, "Before the end of the war the latest new developments needed to totally diverge from those of the previous types." This applied particularly to completely new aircraft types, whose flight performance was to be substantially increased. Preliminary planning had begun as early as the beginning of 1916; however, actual flight testing took place in 1917 and 1918.

Intended for the long-range reconnaissance role was an optimized, extremely light and aerodynamic biplane layout, with *Idflieg* (*Inspektion der Fliegertruppen* = Inspectorate of Flying Troops) requirements of a minimum ceiling of at least 5,000 meters and a maximum speed of at least 165 km/h at that altitude. Preliminary plans for the 7C1 reconnaissance plane and its 7D1 single-seat fighter counterpart envisioned a short fuselage with an oval cross section, single-bay wing surfaces that organically flowed from the fuselage, small rudder without fixed fin, and smooth, streamlined surfaces. Rumpler had stipulated circular fuselage cross-sections since 1911, but no Rumpler series

Right: Edmund Rumpler in front of the fuselage structure of the 7C1 during the fabric band wrapping process. Two layers of fabric were wrapped diagonally around the framework in opposite directions.

Below: Load test for the horizontal stabilizer of the Rumpler 7C1.

production aircraft had used it yet. Experience with the strength and rigidity of single-bay wings was first provided to the Rumpler design team by their 5DDD biplane followed by the Rumpler 6A2 prototype aircraft.

At the end of November 1916 a mockup of the front fuselage section for the Rumpler 7C1 and 7D1 was completed for demonstration purposes. The factory mockup exhibited extremely lightweight construction and uncompromising effort to achieve the lowest aerodynamic drag – characteristics later displayed in the postwar Rumpler *Tropfen-Auto* streamlined automobile. The fuselage had an oval cross section with extremely thin stringers. Separate upper wing panels bent downward at the wing root in gull-wing configuration to rest directly over the completely enclosed 160 hp Mercedes D.III engine. The engine was mounted within the fuselage under the integrated cabane structure, which also included the radiator. For this design method Rumpler obtained a patent in 1919, which noted "with mid-wing monoplane aircraft the fuselage enclosing the engine is used for attaching the upper wing."

In December 1916 *Idflieg* contracted Rumpler to build three prototypes. In January 1917 these three prototypes, the two-seat 7C1 and the 7D1 and 7D2 single-seat biplanes, were under construction. The wings of all three aircraft were braced with a single I-strut. The fuselage nose started with a large propeller spinner blended into the fuselage contours. The plywood-covered empennage was entirely cantilevered and without fixed fin to save weight. According to a patent issued in early 1915 the fuselage received a double-layer covering of Cellon-soaked fabric, diagonally wrapped, that was extremely light and resistant to warping.

During the flight tests, which took place from March through August 1917, the prototypes demonstrated a top-speed of 170 km/h, noteworthy considering the 160 hp Mercedes D.III engine used. The flight tests also showed, however, that the sophisticated aerodynamics detracted from the handling of the aircraft. The broad I-Struts and the cabane structure integrated into the fuselage obstructed the pilot's view. The cabane structure also negatively affected the flow of air from the propeller, so that the engine did not achieve its full revolutions, and thus its full power.

There were other practical difficulties. During installation the engine had to be slid in from the front, and during routine maintenance it was nearly inaccessible. In addition, test pilot Friedrich Budig determined during the first flight with the 7C1 that the rudder and ailerons worked irregularly. Aileron ineffectiveness was largely due to use of a worm-gear

Above: The sleek, advanced lines of the first version of the Rumpler 7C1 are shown to advantage in this snowy scene taken in January 1917. Already fitted with its 160 hp Mercedes D.III engine, the 7C1 is shown awaiting wing installation.

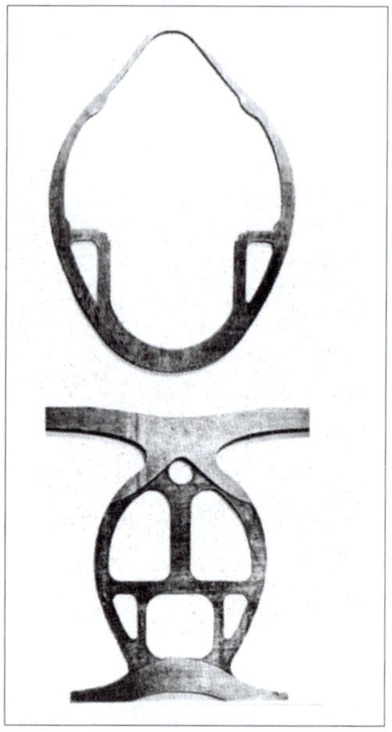

Above: The Rumpler 7C1 fuselage formers were elaborately shaped.

actuating mechanism, rare among aircraft designs.

Moreover, directional stability was inadequate due to the short fuselage and lack of a fixed fin. After attachment of a fixed fin brought insufficient improvement, the fuselage was lengthened. Pitch stability was improved because the tailplane had an inverted wing profile but was still insufficient, particularly when the airplane lost speed, which caused the nose to pitch up. Lateral stability was sufficient, but the limited aileron effectiveness gave marginal control and a slow roll rate. The effect of the ailerons improved somewhat after the outer portion of the aileron washout was bent downward. Finally, the aircraft were plagued with in-flight vibration and twisting of the fuselage, clear indications that the light-weight structure was not robust enough for combat operations.

Even before flight testing was completed, development of a two-seat fighter based on the 7C1 was considered. Contrary to initial planning, *Idflieg* communicated on 15 May 1917 that it required a "non-bomber reconnaissance aircraft as a replacement for the 7C1." Rumpler proposed that altitude tests should be undertaken as soon as possible, which were to be conducted by test pilot Gustav Basser. Among the structural alterations considered it was determined that the aircraft should be "an X-Strutter like the 6A2." However, despite this suggestion, the I-struts remained in place.

Despite a reference[1] stating the Rumpler 7C1 saw limited production as a batch of 20 aircraft designated Rumpler C.IX, this is highly unlikely.

Rumpler 7C1 Specifications	
Engine:	160 hp Mercedes D.III

First, no photos have been found of such an aircraft in service; such photos surely would have been taken given the striking appearance of the 7C1. Second, the late Peter M. Grosz, dean of WWI German aircraft historians, stated that the Rumpler C.IX was in fact based on the Rumpler C.IV and C.VII airframe, with power from a 245 hp Maybach Mb.IVa engine and equipped with a long-range fuel tank[2]. Third, the long, troubled, and ultimately unsuccessful development of the companion Rumpler 7D1 fighter design into the Rumpler D.I has also been well-documented by Peter M. Grosz[3]. Compared to the extensive information and photographs of that protracted development, the trail of the 7C1 stops with a photograph of the attractive last version of the design with large fixed fin and test pilot Gustav Basser in the cockpit (see page 116).

In Retrospect

The extensive development difficulties that prevented the Rumpler D.I from reaching operations despite intense engineering work apparently derailed the Rumpler 7C1 even earlier in its development in favor of the more practical Rumpler C.X. Rather than totally diverging from existing types as Edmund Rumpler favored, the successful C.X was an evolutionary design. In contrast, the unsuccessful 7C1 diverged too far from existing types.

The 7C1 also suffered from a common malady of

Above & Below: The completed Rumpler 7C1 prototype in January 1917 shows its elegant streamlining. The I-struts and tail unit without fin look too fragile for the harsh demands of front-line operational service.

1. Kranzhoff, Jörg A., *Edmund Rumpler – Wegbereiter der industriellen Flugzeugfertigung*, Bernard & Graefe Verlag, Bonn, 2004.
2. Peter M. Grosz, *Rumpler C.IV*, Albatros Publications, Ltd., Berkhamsted, UK, 1992.
3. Peter M. Grosz, "The Rumpler D.I "Super-Plane"? – The Fable and the Facts", *Over the Front* Vol. 4 Number 1, Spring 1989.

Above: The prototype Rumpler 7C1 after a fixed fin was added to improve directional stability looks much more convincing than the initial configuration without fixed fin, February 1917. Although the fin improved directional stability, the fuselage also had to be lengthened to achieve satisfactory results. Pitch stability was improved by the lengthened fuselage but was still insufficient; pitch-up was experienced when the aircraft lost speed.

Above: Test pilot Gustav Basser in the cockpit of the elegant Rumpler 7C1 after a fixed fin was added to the tail, February 1917. Although lateral stability was satisfactory, aileron effectiveness was marginal and the aircraft had a slow roll rate. In addition, the aircraft was plagued with in-flight vibration and twisting of the fuselage, clear indications that the light-weight structure was not robust enough for combat operations.

German and Austro-Hungarian aircraft designs in both world wars; trying to save weight with a tail that was too small and a fuselage that was too short. The result was an unstable design that had to be modified with larger tail surfaces and longer fuselage for safe flight characteristics.

Moreover, by mid 1917 it was certainly realized that, for performance competitive with powerful late-war fighters, a photographic two-seater would require a much more powerful, and thus larger and heavier, engine than the 160 hp Mercedes D.III in the 7C1. In fact, this should have been obvious in 1916 when the 7C1 was being designed because Rumpler itself was building the more powerful C.III and C.IV. This lack of foresight made a complete redesign necessary and, with the intractable D.I undergoing development along with other types, Rumpler did not have enough engineering resources for that.

The delicate, highly aerodynamic design of the Rumpler 7C1 was more appropriate to a racing or record-breaking aircraft than a robust, reliable front-line combat aircraft.

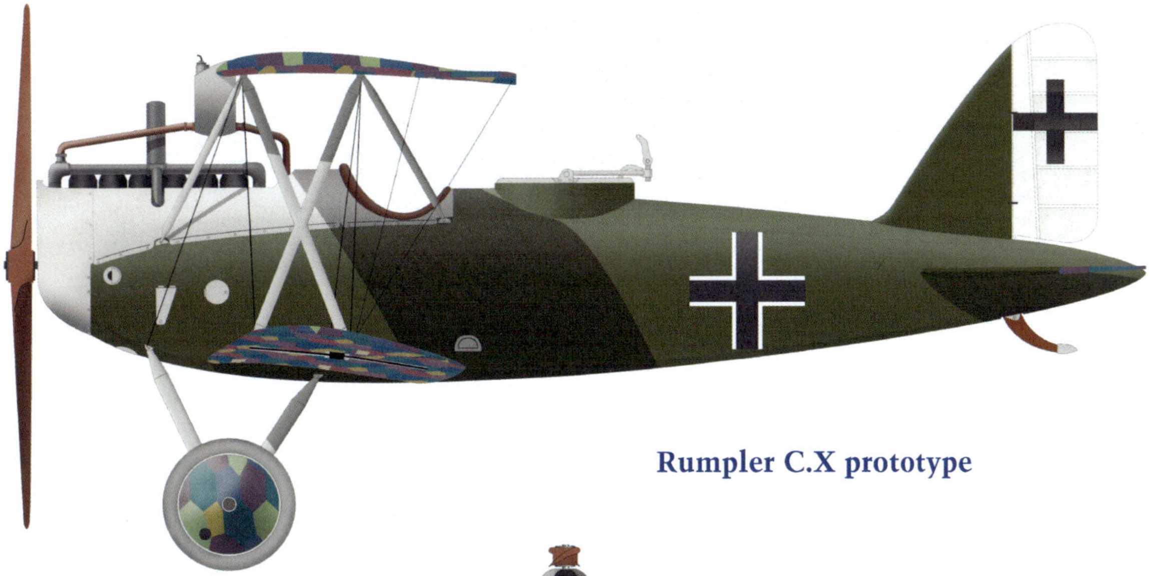

Rumpler C.X prototype

Rumpler C.X

Above: The handsome Rumpler C.X was developed from the earlier Rumpler designs. Its more compact, single-bay design gave better speed, climb rate, and ceiling than earlier Rumplers but it entered production too late to see service.

The Rumpler C.X was the last production type in a long line of successful Rumpler two-seaters. The Rumpler C.I, powered by the 160 hp Mercedes D.III engine and its contemporaries, entered service in 1915 and was an immediate success. It was kept in production for the remainder of the war, initially as a combat aircraft, later as a trainer. The C.III, powered by the 200 hp Benz Bz.IV engine, entered service in 1916, only to be quickly superceded by the even more powerful and successful C.IV.

Rumpler continued development of C-type aircraft throughout the war. While more radical designs like the 7C1 were explored, all production aircraft evolved from earlier designs and all later operational types used the basic C.IV airframe. At the beginning of 1918, Rumpler tested the prototype 8C12, with 260 hp Daimler D.IVa engine and the wing cellule of the C.IV, but with two 'U'-profile interplane struts. Upon test flying the 8C12, it was determined that it had an unsatisfactory climb performance. Test pilot Fredrich Budig was surprised at these results as, when compared to the experimental 6A7 aircraft (prototype C.IV), the 8C12 prototype had been computed as having an additional 57 kg of propeller thrust. Due to the reduced drag it was calculated that the required climb should have been possible with 230 hp.

Since the propeller efficiency was obviously not reduced, Budig decided to introduce measures to make the aircraft more aerodynamic. Like the prototype C.I, he decided to change the angle of attack of the wings. The result was that the aircraft's climbing performance improved by about one minute.

The prototype 8C13 received likewise the 260 hp Daimler D.IVa engine, but used the wing cellule of the C.VII. From this aircraft the prototype 8C14 was developed in early 1918 that became the series version of the Rumpler C.X. Its 260 hp Maybach Mb.IVa engine was completely enclosed in an aluminum cowling. The compact, almost oval fuselage was only 6.9 meters long and well streamlined. The wings were staggered and connected by 'X' struts, which tilted outward at the top. The wing spars, manufactured from pine, had a double 'T' cross-section. In contrast to the lower wing, the upper wing had neither sweep nor dihedral, and had a span of 10.50 meters and balanced ailerons. The lower wing span was 9.00 meters. The cantilever tail unit had a short, straight dorsal fin, which was built as an integral part of the fuselage. The rudder was relatively narrow and had a right-angled vertical trailing edge. The horizontal tail unit had the proven negative airfoil curvature.

In the summer of 1918 the C.X prototype was damaged in a crash landing. Despite this setback, the test program was not affected. With an empty weight of 950 kg the aircraft was lighter than the earlier C.IV. The smaller wing and single-bay bracing were responsible for much of the weight saving and also reduced drag. With the over-compressed 260 hp Maybach engine coupled with a lighter, more aerodynamic airframe than the C.IV, the C.X had a maximum speed of 195 km/h and a service ceiling of 7,800 meters, which it could reach in 59 minutes. Thus the C.X was by far the highest flying and fastest German C-type aircraft of the war. "Based on this performance we now received an order," said Gustav Basser in August 1918. However, only a few examples were built and tested by the end of the war.

Rumpler C.X Specifications

Engine:	260 hp Maybach Mb.IVa	
Wing:	Span (Upper)	10.50 m
	Span (lower)	9.00 m
	Chord (Upper)	1.80 m
	Chord (Lower, mid-wing)	1.40 m
	Area	29.0 m^2
General:	Length	6.90 m
	Empty Weight	957 kg
	Useful Load	435 kg
	Loaded Weight	1,385 kg
Maximum Speed:		195 kmh
Climb:	6000m	27.0 min
	7800m	59.3 min
Service Ceiling:		7800m
Range:		585 km

The Rumpler C.X had the performance to carry on the Rumpler tradition of being the leading German long-range reconnaissance aircraft, but the war ended before it could reach operations.

Below: The simplified struts and wire bracing of the Rumpler C.X compared to the earlier, two-bay Rumpler reconnaissance planes using the C.IV airframe gave reduced drag and higher performance.

Left: Sometime in the summer of 1918 the Rumpler C.X prototype suffered damage during a forced landing, but this did not delay development.

This Page: The Rumpler 8C12 prototype eventually led to the 8C14 design that was given the military designation Rumpler C.X. Whereas the production C.X clearly used a new airframe, the 8C12 prototype appears to have used an airframe from a standard Rumpler C.IV as illustrated by the flat-sided fuselage and C.IV tail surfaces. However, the wings were modified with U-struts for interplane bracing similar to those on the Rumpler 7D7 fighter prototype.

Rumpler Fighters

Fighters were a sideline for Rumpler, but one that grew increasingly important in the last two years of the war.

The Rumpler 6A2 was a two-seat fighter that was a refined, single-bay derivative of the Rumpler C.I reconnaissance airplane. It remained a prototype.

Ironically, the most successful Rumpler fighter was the Rumpler 6B1 floatplane derived from the C.I two-seat reconnaissance airplane. Despite being developed from a two-seater, the Rumpler 6B1 was as successful on operations as any of the Navy's other single-seat floatplane fighters. It was followed by the closely-related 6B2 used in small numbers.

The high-altitude Rumpler D.I was painstakingly developed through a long series of prototypes and, due to its potential, even received a production order for 50 aircraft. These became flying development aircraft as Rumpler sought to bring the D.I design to a practical production fighter. But despite the intense effort, prolonged development, and small production order, Rumpler was unable to deliver a practical fighter before the Armistice. Without question the D.I saga was Rumpler's greatest fighter effort and greatest disappointment.

Rumpler 6A2

Above: The Rumpler 6A2 first flown in January 1916 was a two-seat fighter. The first version used 'X'-struts. (The Peter M. Bowers Collection/The Museum of Flight)

The single-bay 6A2 was designed as a two-seat fighter. Powered by a 160 hp Mercedes D.III, its first test flight in January 1916 revealed problematic handling qualities, but extensive experimentation and modification finally gave the 6A2 'excellent flight characteristics' according to Friedrich Budig.

Rumpler 6A2 Specifications		
Engine:	160 hp Mercedes D.III	
Wing:	Span	10.20 m
	Area	36.00 m^2
General:	Empty Weight	780 kg
	Loaded Weight	1260 kg
Climb:	1000m	6.0 min
	2000m	16.5 min
	3000m	30.7 min
	4000m	69.0 min
Ceiling:		4200m

Unfortunately, climb rate was less than the C.I reconnaissance airplane and the wings were too 'elastic', so the X-struts were replaced with parallel struts. The 220 hp Mercedes D.IV straight-eight engine was also tried and gave very good climb rate but vibration was excessive. Records are ambiguous about whether the 160 hp Mercedes D.III was re-installed. Both flight testing and calculations indicated the 6A2 would not meet its requirements, and it was abandoned in favor of the enlarged, two-bay 6A5 prototype that became the C.III of otherwise similar appearance. The intermediate 6A3 and 6A4 were designed but probably not built.

Above & Below: The Rumpler 6A2 was powered by a 160 hp Mercedes D.III. The first version shown here used 'X'-struts.

Above: Rumpler test pilot Gustav Basser at left in front of the Rumpler 6A2 at Johannisthal. Although not a successful design, the 6A2 was instrumental in development of the larger, two-bay 6A5 that became the Rumpler C.III and lead to the Rumpler C.IV. The 6A2 introduced the basic fuselage and tail assembly and even the wing planform that are seen in the later and larger, more powerful two-bay Rumpler C.III reconnaissance airplane.

Above: The second version of the 6A2 used parallel struts for greater torsional strength of the wing cellule giving improved aileron response and better controllability.

Rumpler 6B1 & 6B2

Above: Rumpler 6B1 Marine #1045 of the last 6B1 production batch shows its propeller spinner and streamlined nose in this air-to-air photograph. Production 6B1 aircraft had the Rumpler C.IV's curved horizontal tailplane. Surprisingly, the 6B1 was more effective on operations than either the Albatros W4 or Brandenburg KDW. Pilots liked its excellent handling qualities and its performance was competitive despite being derived from a two-seat reconnaissance airplane.

The Rumpler 6B1 (known to the Navy as the Rumpler ED) was derived from the successful Rumpler C.I two-seat reconnaissance aircraft. The weight of the floats was compensated for by elimination of the observer and his equipment, and the upper wing was moved forward to compensate for the forward shift in center of gravity. Like the Rumpler C.I and Albatros W4, the 6B1 was powered by a 160 hp Mercedes D.III. The 6B1 retained the C.I's single fixed gun for the pilot. The prototype, Marine #751, was delivered to the seaplane testing command on 7 July 1916. After testing it was accepted on 10 August, and the first production batch of ten fighters was ordered on 14 August, followed by a further batch of 25 fighters.

Rumpler 6B2

The Rumpler 6B1was followed in production by the improved 6B2 that had the ability to mount two machine guns. The 6B2 also had refined aerodynamics derived from later production Rumpler C.IV aircraft that gave it greater speed; specifically the spinner was replaced by a rounded nose. Three production batches totaling 50 aircraft were ordered in early 1917. However, production was slow; the first aircraft were delivered a full ten months after being ordered, and only about half were fitted with two guns, the others mounting a single gun. Production data and other information about use suggest that by the time the 6B2 was ready, the Navy strongly preferred two-seat seaplane fighters, reducing the urgency for the 6B2 and other single-seat fighters. The 6B2 therefore saw limited operational service and was mostly used for training. Four 6B2 aircraft were sold to the Austro-Hungarian Navy, which assigned them serial numbers E1–E4.

Above: The first prototype Rumpler 6B1, Marine#751, is photographed here conducting flight trials at the Rumpler facility on the Müggelsee near Berlin in June 1916. The armament has not yet been installed.

Above: The first prototype Rumpler 6B1 while at the *SVK* in Warnemünde. The light color does not contrast well with the sky background, which was good for air-to-air camouflage and bad for photography. Power was provided by a 160 hp Mercedes D.III engine. The first three prototypes had the C.I horizontal stabilizers with straight leading edges; production aircraft had the C.IV tailplane with curved leading edges.

Above: The first prototype Rumpler 6B1, Marine#751, while at the *SVK* in Warnemünde. The single, fixed machine gun for the pilot has now been installed.

Above: The first prototype Rumpler 6B1, Marine#751, while at the *SVK* in Warnemünde. This view shows the clean lines of this large floatplane fighter. Being a large, two-bay design derived from a reconnaissance two-seater, the 6B1 was not as fast as the competing Albatros W4 and Brandenburg KDW. However, it had a reasonably good performance and excellent handling and flying qualities; on operations it was as successful as the Albatros W4 and far more successful than the fast but tricky Brandenburg KDW.

Above: The prototype Rumpler 6B1 photographed at the *SVK* at Warnemünde, where the aircraft was delivered on July 7, 1916. It was accepted on August 10, 1916, and this was quickly followed by a production order for the first batch of 10 aircraft on August 14, before the next two prototypes, 787 and 788, were completed. Someone has drawn in the outline of the fin and rudder on the original print due to the low contrast between the sky and the aircraft.

Below: Rumpler 6B1 W4 Marine #751 is shown after being repainted in camouflage colors after assignment to Zeebrugge. The straight leading edge of the C.I-style tailplane is clearly shown. Flying from Zeebrugge, *Lt.z.S.* Neimeyer used Rumpler 6B1 #751 to down a Short Seaplane on 31 Aug. 1916 and a Caudron G.4 on 7 Sept. 1916.

Rumpler 6B1 & 6B2 Production

Marine Numbers	Qty	Notes
751, 787–788	3	Prototypes, C.I tailplane
890–899	10	First 6B1 production, C.IV tailplane
1037–1061	25	Second 6B1 production
1062–1066	5	6B2 pre-production
1188–1207	20	First 6B2 production
1434–1458	25	Second 6B2 production

Above & Below: Rumpler 6B1 Marine #1045 photographed in flight. This fighter was stationed at Borkum on September 13, 1917, where these photos and the one starting this section were probably taken. The photo below shows the camouflage on the upper surfaces.

Left: Rumpler 6B1 Marine #1045 displays its agility for the photographer.

Above & Below: Move views of Rumpler 6B1 Marine #1045 performing in flight.

Above & Right: Move views of Rumpler 6B1 Marine #1045 performing in flight.

Below: Rumpler 6B1 after landing, probably Marine#1045.

Above: Rumpler 6B1 Marine #1045 in flight. All 6B1 fighters except the three prototypes had the curved C.IV horizontal tail.

Below: Rumpler 6B1 Marine #788, the third prototype.

The prototype Rumpler 6B1 after armament was fitted

Rumpler 6B1 Marine #751 after camouflage was roughly applied at Zeebrugge. This is how the aircraft looked when flown on operations.

Rumpler 6B1 Marine #1042 in standard late-war naval camouflage.

Above: Rumpler 6B1 Marine #899 was the last aircraft of the first production batch. It is seen here being hoisted by a crane for launching, something every German seaplane was equipped to do.

Below: Rumpler 6B1 Marine #899 after successful launch. The curved tailplane applied to production aircraft is clearly visible. The standard late-war, three-color lozenge fabric does not appear to have been used on this aircraft; instead, the upper surfaces of the wings appear to be sprayed in two colors. On 15 May 1917 #899 was stationed at Borkum on the Flanders coast for operations over the North Sea.

Above: Rumpler 6B1 W4 Marine #1051 shows its dark coloring and operational markings. The 6B1 and 6B2 had good performance despite their size and had far better handling qualities than the competing Brandenburg KDW.

Above: Rumpler 6B1 Marine #1042 of the second production batch is launched into the water and is seen just leaving the launching dolly. This aircraft was assigned to Konstanza on May 15, 1917, suggesting this photograph was taken there.

Rumpler 6B1 Marine #1051 of the second 6B1 production batch

Above: Rumpler 6B1 Marine #1059 serves as a backdrop for a group photograph; unfortunately, the identities of the men and dogs are not known. Either this is a very informal photograph or the photographer made this shot while setting up for the formal portrait. The Marine Number has been painted in the side of both floats to identify the aircraft, and the Axial propeller has a metal tip to reduce erosion from spray. Aviators seem to like dogs.

Above: Rumpler 6B2 Marine #1066 was from the 6B2 pre-production batch. From the rear it is difficult to tell the 6B1 from the 6B2. Although the 6B2 was designed to carry two machine guns, only about half the aircraft actually had two guns.

Above: Rumpler 6B2 floatplanes 1439, 1062, 1063, and 1194 are lined up at either Kiel Holtenau or further east at one of the Baltic *Seeflugstationen* (naval air stations). Marine #1439 carries a large white circle as a tactical identification marking, while #1062 carries two white circles and #1063 carries three white circles.

Right: This Rumpler 6B1 has been repainted in the late, straight- sided national insignia specified by the Navy on March 30, 1918. It is being readied for a training flight at Putzig naval air station.

Rumpler 6B2 Marine #1063 in standard late-war naval camouflage and markings with three white circles added as a tactical marking.

Rumpler 6B2 Marine #1066 of the pre-production batch of 6B2 fighters. By this time the standard late-war naval three-color camouflage fabric was standard on the upper surfaces.

Rumpler 6B2 Marine #1439 in standard late-war naval camouflage with a large white circle added as a tactical marking. The fin also appears to be in three-color naval camouflage fabric, perhaps resulting from a repair.

Above: Rumpler 6B2 Marine #1189 was from the first 6B2 production batch. Its lack of propeller spinner was due to research to improve the Rumpler C.IV series that showed a streamlined nose without spinner had less drag. This made the 6B2 14 km/h faster than the similarly-powered 6B1, a significant improvement for such an apparently minor change. Here is is shown on its beaching gear at *Seeflugstation Puntisella* on July 9, 1918, in Austro-Hungarian service.

Left: Rumpler 6B2 in postwar Dutch service.

Right: An Austro-Hungarian Rumpler 6B2 flies over the harbor at Trieste.

Above & Below: Austro-Hungarian Rumpler 6B2 E3 (formerly Marine #1090) is moored next to the officers' mess at Igalo naval air station. The upper surfaces of the wings are in the German naval three-color printed camouflage fabric, but unusually there are no national insignia on the upper wings.

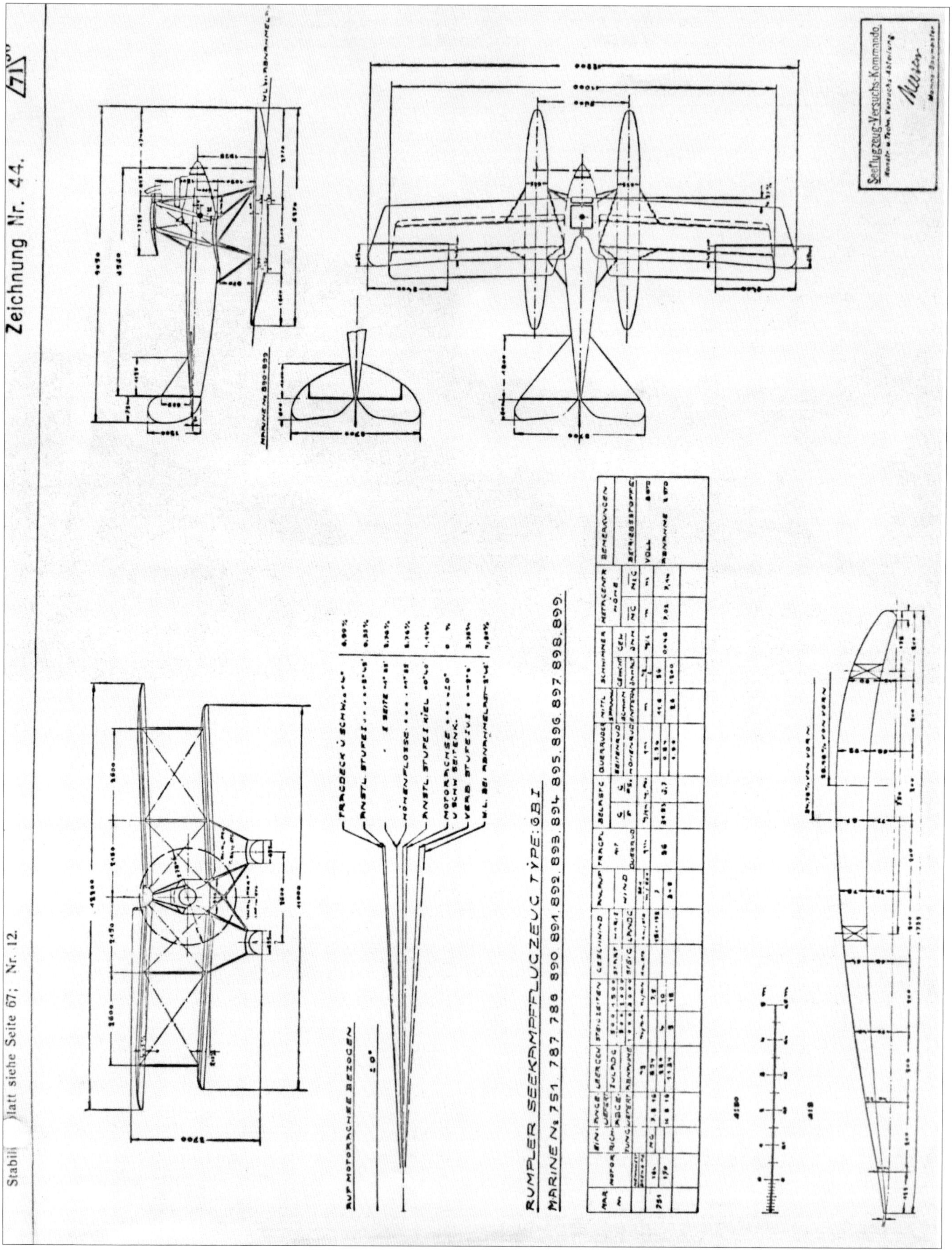
Zeichnung Nr. 44.
Stabil[illegible] Blatt siehe Seite 67; Nr. 12.
RUMPLER SEEKAMPFFLUGZEUG TYPE: 6B I
MARINE No 751 787 788 890 891 892 893 894 895 896 897 898 899.
Seeflugzeug-Versuchs-Kommando

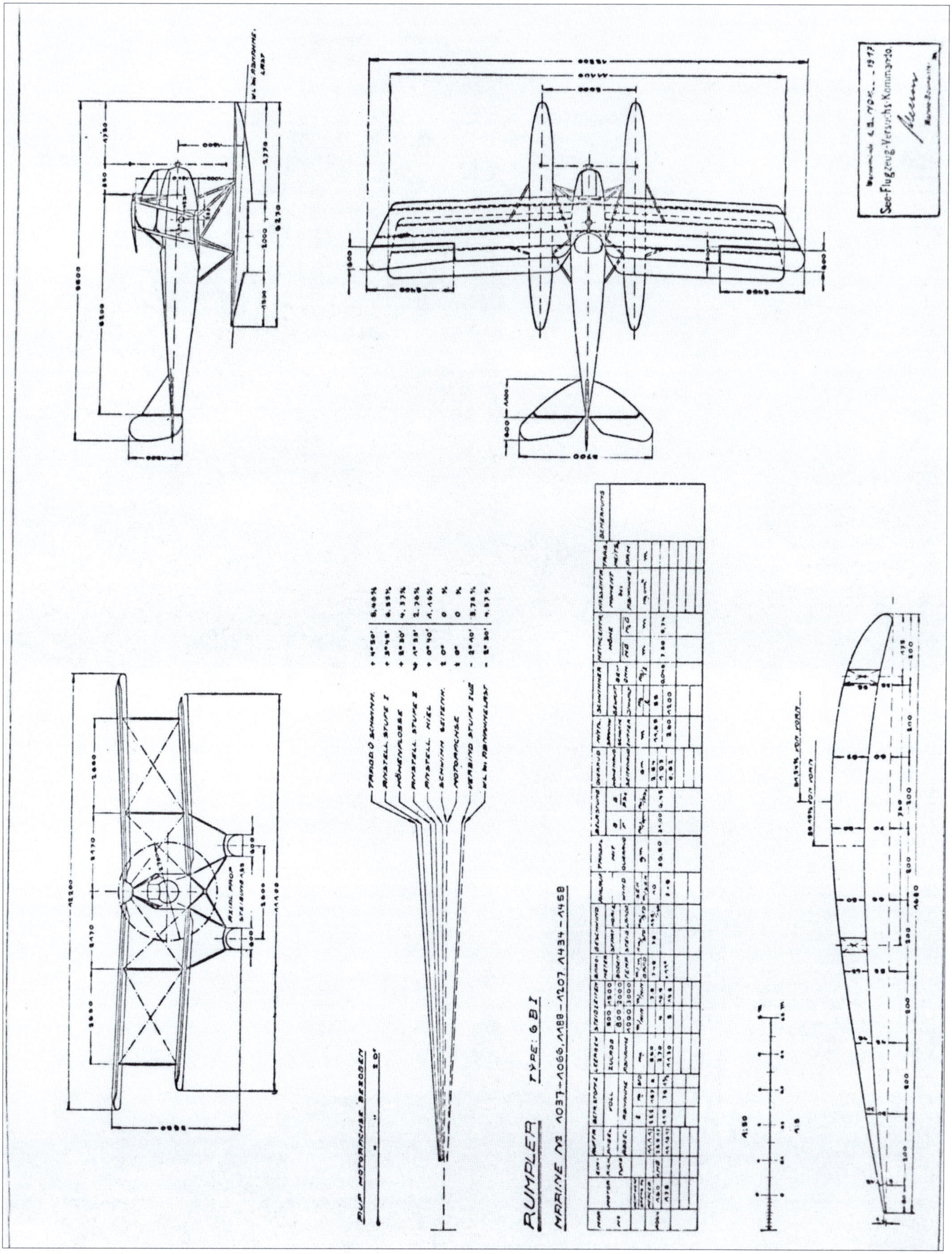
RUMPLER TYPE: 6 B I
MARINE No 1037-1066, 1188-1207, 1434-1458
See-Flugzeug-Versuchs-Kommando

Rumpler D.I

Above: Two Rumpler D.I fighter prototypes were entered in the Second Fighter Competition. Both were powered by the Mercedes D.IIIa engine. D.I 1589/18 above was photographed postwar so the guns have been removed.

The story of the prolonged development of Rumpler's only production land-based fighter, the D.I, and all the prototypes that led up to it is a story of great vision and innovative engineering that ultimately failed to produce a practical fighter despite intensive effort.

Design studies for the Rumpler single-seat fighter began in early 1916 in parallel with its two-seat, high-altitude reconnaissance counterpart, the 7C1. Like the 7C1 the fighter prototype, known internally as the 7D1, was based on extensive investigation into new aerodynamic features supported by new structural techniques. To achieve the desired climb, ceiling, and speed at altitude the aircraft had to be very light and streamlined. As a result the 7C1 and 7D1 prototypes resembled aircraft designed for breaking speed and altitude records and did not have the flying qualities and robust structures needed for practical combat aircraft.

By November 1916 Rumpler was able to build a mock-up of the 7D1 and the next month *Idflieg* ordered three prototypes of the fighter. Like the 7C1, these were powered by the 160 hp Mercedes D.III, then the best available fighter engine. *Idflieg* required a climb to 5000 m in 33 minutes and a maximum speed of 165 km/h at that altitude, and also required installation of supplementary oxygen for the pilot, proof that exceptional climb and ceiling were planned.

Rumpler completed the 7D1 prototype (and an 'identical' 7D2 which is otherwise unknown) in spring 1917. The 7D1 was a highly-streamlined biplane with top wing supported by the engine cowling. It was built with a series of light-weight fuselage stringers covered with doped fabric for streamlining and light weight. Single I-struts on each side braced the wings and only two bracing wires were used for each bay.

Flight test were disappointing. Not only did the engine cowling obscure the pilot's forward vision, a serious problem for any aircraft and especially for a fighter, but the airflow interference between the

Above: Mockup of the forward fuselage section of the Rumpler 7D1 in November 1916 shows the interesting structure and integration of the engine into the cabane structure. The difficult engine installation and restricted maintenance access are evident.

Above: Fuselage construction of the Rumpler 7D1 on February 6, 1917. The stringers were covered with two layers of fabric strips applied diagonally in opposite directions after being soaked in aircraft dope. This method was patented by Rumpler in 1915 and was an attempt to blend strength with minimum weight.

Right & Below: The Rumpler 7D1 prototype at Johannisthal in the spring of 1917 shows great attention to streamlining. Unfortunately, the engine cowling not only obscured the pilot's forward vision, it also interferred with the airflow from the propeller, preventing the engine from achieving its full RPM and thus its full power.

propeller and the cowling and upper wing structure prevented the engine from developing its full RPM and thus its full power. Moreover, the buried engine was virtually inaccessible for maintenance.

As a result, Rumpler designed a revised prototype with conventional center-section, the 7D3 (according to chief test pilot Friedrich Budig). The 7D3 was probably not successful because little is known about it, and the following 7D4 prototypes were much more widely tested. The first 7D4 prototype, powered by a 160 hp Mercedes D.III, was flight-tested in October 1917 and demonstrated a climb to 5,200 m in 20 minutes, an exceptional rate of climb for a fighter powered by a water-cooled, in-line engine.

Two different 7D4 prototypes competed at the First Fighter Competition in January 1918, two years after start of development. The U-strut 7D4 prototype climbed to 5,000 m in 23.8 minutes, and the parallel-strut 7D4 climbed to the same altitude in 28.0 minutes. The significantly slower climb of the parallel-strut 7D4 motivated the development team to focus on the U-strut configuration for the production fighter. By this time *Idflieg* had designated the Rumpler fighter as the Rumpler D.I regardless of the details of internal company designations. Pilots at the competition were impressed by the Rumpler fighter's climb but reported it was excessively touchy and difficult to fly.

According to engineer Theo Rockenfeller the Rumpler 7D5 was a fighter fitted with an automobile-type radiator in the nose that was built and tested; unfortunately, nothing else is known about it. The 7D6 was a cantilever quadraplane fighter design of late 1917 that fortunately was not built. The next Rumpler prototype fighter to fly was thus the 7D7, which appeared too late to compete at the First Fighter Competition.

Rumpler 7D7 Specifications

Engine:	170 hp Mercedes D.IIIa	
Wing:	Span Upper	8.42 m
	Span Lower	7.38 m
	Chord Upper	1.32 m
	Chord Lower	1.09 m
	Gap	1.365 m
	Area	16.66 m²
General:	Length	5.75 m
	Height	2.56 m
	Empty Weight	615 kg
	Loaded Weight	805 kg
Maximum Speed (at 5000 m):		180 km/h
Climb:	5000m	23.7 min

The 7D7 design was modified from the U-strut 7D4 by using a new airfoil and smaller control surfaces. The smaller cockpit was moved forward 35 cm and the inefficient airfoil radiator used in the 7D4 prototypes was replaced by a pair of ear radiators attached to the fuselage above the lower wing. The new radiators initially were not effective and significant engineering work had to be done to fix the problem. The improved 170 hp Mercedes D.IIIa engine was used, and flight tests showed the 7D7 was faster than the 7D4 and demonstrated a climb to 5,000 m in 18 minutes, a significant improvement over an already impressive climb rate.

The Fokker D.VII won the First Fighter Competition and went on to win fame over the Western Front. The Rumpler D.I, now based on the 7D7 prototype, was viewed as promising enough that *Idflieg* ordered an evaluation batch of 50 aircraft, something that was also done for the Roland D.VI

Left: The Rumpler 7D3 prototype was photographed in December 1918 at the Breslau Industrial Fair, for which it was repainted. Built in mid-1917, it had a conventional center section and parallel interplane struts. The cut-away skin on the fuselage reveals the original construction method was retained.

Above & Below: Two Rumpler 7D4 fighter prototypes were tested at the First Fighter Competition; the parallel strut version above and the U-strut version below. Both were powered by the Mercedes D.IIIa engine. Based on further research applied to the Rumpler C.IV, these had no propeller spinner. The nose and tail of these aircraft are now covered in thin plywood for increased strength. The photo above was taken 24 January 1918 at the start of the First Fighter Competition. (Lower photo from The Peter M. Bowers Collection/The Museum of Flight)

and SSW D.III. The Rumpler D.I production batch was given serials D.1550/18 to D.1599/18.

The D.I, based on the 7D7, underwent *Idflieg's* standard test and evaluations procedures starting in February 1918. Load tests lasted from February 22 to May 1 with a number of failures and fixes during that time. The summary report of May 18 called for additional reinforcement of various items, indicating that the Rumpler team had left little margin for error when designing the aircraft for minimum weight. The light-weight fuselage of the 7D7 twisted during flight maneuvers, and flight tests in April 1918 revealed problematic flight characteristics. *Idflieg* stated that the type was "unacceptable for the front and would be rebuilt."

To solve the fuselage twisting problem and

Above & Below: Two more views of the Rumpler 7D4 fighter prototypes that were tested at the First Fighter Competition. The U-strut version is above and the parallel strut version is below. Both were powered by the Mercedes D.IIIa engine. The U-strut version was selected for further development. The 7D4 prototypes had an aerodynamic balance for the rudder. (Lower photo from The Peter M. Bowers Collection/The Museum of Flight)

improve flight characteristics, a completely new, stronger fuselage design had to be used for the production fighters. The revised design, based on a fuselage wrapped diagonally with strips of plywood similar to the *Wimpelrumpf* technique developed by LFG Roland, was known internally as the 8D1, and it became the basis for the production Rumpler D.I.

Three Rumpler D.I production fighters based on the 8D1 design competed at the Second Fighter Competition in May–June 1918. Two of these,

Above: The U-strut version of the Rumpler 7D4 fighter prototype at the First Fighter Competition. *Oblt*. Hermann Göring is thought to be the pilot in the cockpit.

Above: The Rumpler 7D7 fighter prototype was developed from the U-strut version of the 7D4. The rudder now has no aerodynamic balance.

Left & Below: The Rumpler 7D7 prototype was powered by a 170 hp Mercedes D.IIIa engine. None of its controls were aerodynamically balanced. Production D.I fighters normally had horn aerodynamic balances.

Above & Above Right: These front views show the ear radiators mounted above the lower wing on the Rumpler 7D7 prototype to advantage. The earlier 7D4 prototypes had airfoil radiators that were not satisfactory.

Above: The Rumpler 7D7 prototype with anemometer airspeed indicator mounted on the left interplane strut. The 7D7 retained the innovative Rumpler wrapped fuselage which proved insufficiently robust during flight evaluations.

Rumpler D.I (8D1) Specifications

Engine:	185 hp BMW.IIIa	
Wing:	Span Upper	8.42 m
	Span Lower	7.38 m
	Chord Upper	1.32 m
	Chord Lower	1.09 m
	Gap	1.365 m
	Area	16.66 m²
General:	Length	5.75 m
	Height	2.56 m
	Empty Weight	615 kg
	Loaded Weight	805 kg
Maximum Speed (at 5000 m):		180 km/h
Climb:	5000m	13.2 min
	6000m	17.8 min
	7000m	26.9 min
	8000m	53.3 min
Ceiling:		8200m

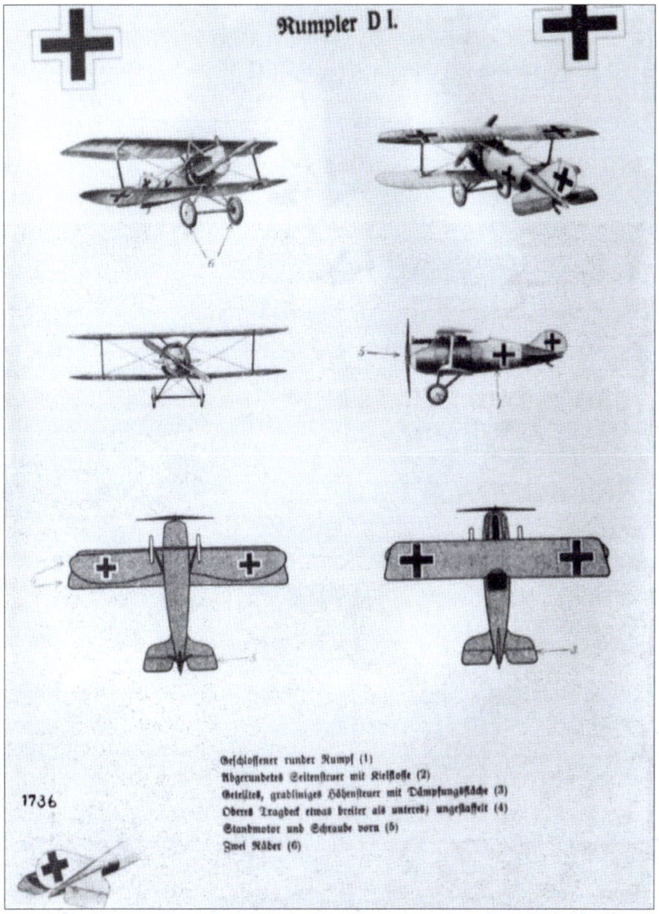

Right: A formal aircraft identification chart was created for the Rumpler D.I fighter, a clear indication that operational service was intended. The chart shows the 7D7 prototype because all control surfaces are unbalanced; the production Rumpler D.I (factory designation 8D1) had horn-balanced ailerons.

Above & Below: Rumpler D.I 1553/18 (w/n 4402, factory designation 8D1) competed in the Second Fighter Competition. Powered by a 170 hp Mercedes D.IIIa engine, it lacked the aerodynamic balances on the ailerons seen on other production D.I fighters. With the 8D1 Rumpler finally abandoned the wrapped fuselage construction it had developed in favor of a stronger, but heavier, fuselage with the fabric replaced by plywood to eliminate tail twisting in flight.

D.1552/18 and D.1553/18, were powered by the over-compressed 180 hp Mercedes D.IIIaü high-altitude engine. Flown by Gustav Basser, these climbed to 5,000 m in 18.7 minutes, an excellent performance. Despite the excellent climb rate, the front-line pilots who tested the Rumpler D.I agreed that it was wholly unsuitable for combat use; it did not glide well, oscillated in flight, and the flying surfaces vibrated and made the pilots feel insecure about the aircraft's structural strength.

The aileron control came in for particular criticism. In the D.I the linkage between the pilot's stick and the ailerons was a worm-gear transmission that turned control rods in the wing. It was light and created no drag, but the resulting aileron control was slow and erratic. Such a control system might be acceptable in a racing or record-breaking aircraft but not in a fighter; fast, consistent roll control is essential to a fighter in combat. Disappointed, *Idflieg* decided that testing and evaluation should continue

This Page: Rumpler D.I 1581/18 (w/n 4430) was the Rumpler entry in the Third Fighter Competition. To compete in the competition, which was for fighters powered by the BMW.IIIa engine, it was fitted with a BMW engine. All other Rumpler fighters were fitted with Mercedes engines. The bump under the fuselage may have covered a liquid-oxygen breathing system.

Above: Rumpler D.I D-289 photographed in March 1919 at Münich while being flown in an airshow by Ernst Udet, a famous *Pour le Mérite* ace credited with 62 victories. Rumpler had given D.I fighters to 28-victory ace Robert *Ritter* von Greim and Enrst Udet after the war for flight demonstations. Udet's D.I was all red with "Rumpler" on the fuselage, probably in black. The "D" on the rudder was black, with "289" a lighter color. (The Peter M. Bowers Collection/The Museum of Flight)

but mass production was out of the question until the aircraft was improved.

A parallel *Idflieg* flight evaluation on production airframe D.1550/18 was summarized on September 9, 1918 and stated that the D.I fitted with aerodynamically-balanced ailerons (the 8D1) flew much better than its predecessor (7D7) with unbalanced ailerons. Maneuverability was very good and most in-flight vibrations were eradicated, but the aircraft was not easy to land and roll-rate was still slow.

Moreover, *Idflieg* identified 42 installation faults

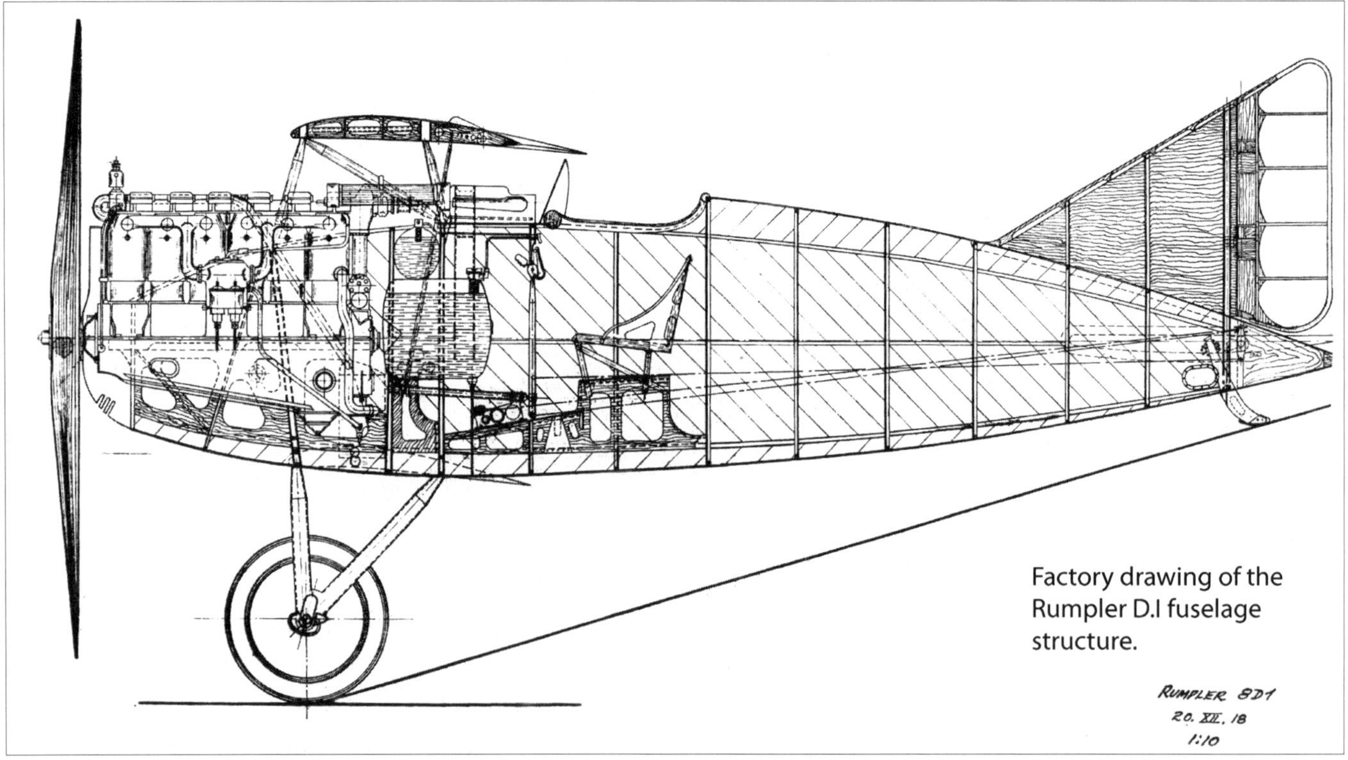

Factory drawing of the Rumpler D.I fuselage structure.

Above: Ernst Udet (left) and Edmund Rumpler in front of Rumpler D.I 1589/18 postwar

Above: Had the war continued, the Rumpler D.I might have been produced in quantity in 1919 for the German Air Service as a high-altitude fighter – if Rumpler could ever manage to get it to meet acceptance requirements. This aircraft, civil registration D-109, was owned by the Bayerische Rumpler Werke.

Rumpler 7D7 during its type testing at Adlershof, Feb. 1918.

Rumpler 7D7 work number 4299, April 1918.

with the prototype! While most were not serious and fairly easy to remedy, the great number of defects calls the competency of the Rumpler design team, and especially its management, into question. What had the design team been doing since the start of type-testing in February?

During load testing between September 16 and October 2 on production airframe D.1550/18 the airframe failed the vertical nose dive requirement and a 4 mm drag cable had to be installed between the U-strut and the nose to pass the test. But the aileron control mechanism failed again, a critical deficiency in a fighter.

Rumpler D.I 1581/18, powered by a 185 hp BMW. IIIa over-compressed engine, was entered and flown at the Third Fighter Competition. This was the only D.I powered by a BMW engine, which was fitted especially for the Third Fighter Competition that specified use of the BMW. The aircraft reached 7,000 m in 26.9 minutes, 8,000 m in 53.3 minutes, and was the only fighter in the competition to reach 8,200 m. The climb and ceiling were clearly superior and flying qualities were reported as good, and the Rumpler D.I was declared one of the two winners of the competition (the other winner was the Fokker V29 parasol monoplane derivative of the Fokker D.VII). However, the D.I was appraised as of limited use in a dogfight against other fighters because it lost height in turns and spun easily, and the short fuselage limited rudder effectiveness.

In Retrospect

Although the D.I had potential as an interceptor, by the Armistice only 22 production aircraft had been built and the type had *still* not passed military acceptance requirements! Extolled as a 'super plane' by the postwar German aeronautical press, the D.I was never a practical front-line fighter. The basic design was too extreme and despite great effort and the investment of tremendous engineering resources, Rumpler was simply unable to develop the aircraft into an effective, combat-worthy fighter.

Rumpler D.I 1553/18 competed in the Second Fighter Competition.

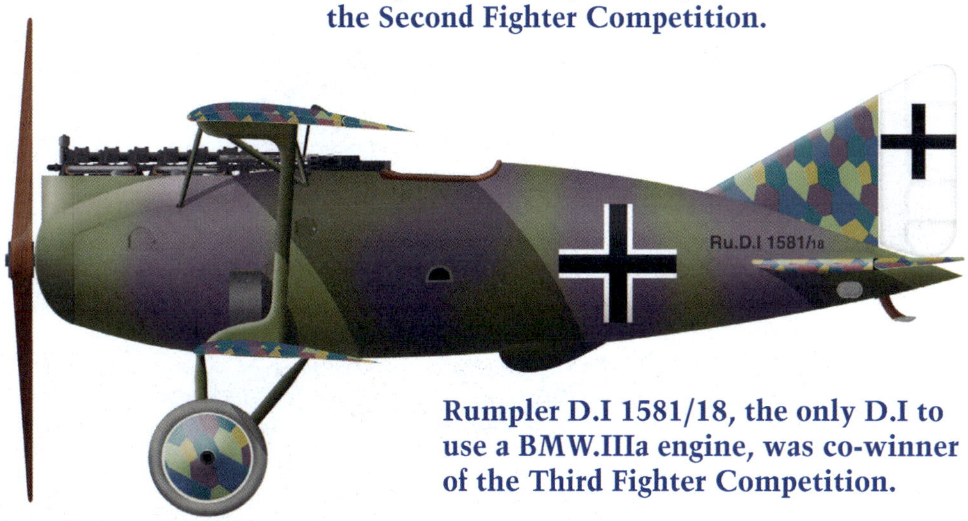

Rumpler D.I 1581/18, the only D.I to use a BMW.IIIa engine, was co-winner of the Third Fighter Competition.

Rumpler D.I D-288 flown postwar by Robert *Ritter* von Greim.

Rumpler D.I D-289 flown postwar by Ernst Udet.

Rumpler Bombers

Rumpler produced a comparatively small number of bomber designs that saw modest production in keeping with their performance. In addition to at least three prototypes, 58 Rumpler production bombers of all types were built as detailed in the adjacent table.

With limited engineering resources, Rumpler eventually abandoned bomber design in favor of more critical fighters and reconnaissance airplanes.

Rumpler Bomber Production

Type	Quantity	Serials
G.I	4	15–18/15
G.II	24	106–117/15 & 119–130/15
G.III	30	300–329/16. See Note.
Note: *Idflieg* ordered 50 Rumpler G.III bombers but only 30 were built.		

Rumpler 4A15

Above: The Rumpler 4A15 was in response to an *Idflieg* request for a *Kampfflugzeuge*, or battleplane. Powered by two 150 hp Benz Bz.III engines mounted as pushers, it set the basic configuration for all subsequent Rumpler bombers.

In July 1914 Rumpler was among several manufacturers asked to develop a *Kampfflugzeuge*, a battleplane or aerial cruiser. The aircraft was to have a crew of two or three with a flexible gun for the observer/gunner mounted in the front cockpit, and was to have at least 200 hp.

Rumpler's response was the 4A15, a twin-engine biplane of conventional wood, wire, and fabric construction powered by 150 hp Benz Bz.III engines mounted in pusher configuration. Each engine was mounted on the lower wing and enclosed in a nacelle housing a frontal Windhoff radiator and a 310 liter fuel tank. A gravity tank was installed above each engine. The 4A15 had a span of 18.75 m and a length of 11.8 m. The simple landing gear included a pair of wheels under each engine and another pair under the

Above & Left: The Rumpler 4A15 was the first Rumpler bomber design, and all subsequent Rumpler bombers followed its general configuration of pusher engines and simple landing gear with nose wheels to prevent nose-overs on landing. The fuel tank below the engine is clearly visible at left. The propellers were mounted on short extension shafts.

nose to prevent nose-overs on landing.

The 4A15 first flew in March 1915 and achieved a maximum speed of 135 km/h at sea level. During informal flight demonstrations the aircraft carried ten people to 3,200 m on March 15th and then carried 16 people to 1,800 m. On 22 March the aircraft reached 2,300 m in 55 minutes while carrying 12 people. However, while flying to Munich on April 10 the 4A15 was forced to make an emergency landing and was damaged. On April 17, 1915 it was destroyed following a carburetor fire. However, during its brief life the 4A15 set the basic configuration for all Rumpler bombers that followed.

Rumpler G.I

Above: This photograph shows the Rumpler G.I with the first design of fin and rudder, the most visible difference between the Rumpler G.I, military designation for 5A15, and 4A15. A gravity tank was also installed just below the upper wing; the main fuel tank for the starboard engine is visible mounted beneath the engine and forward of it. The fuel tank was located near the aircraft center of gravity so burning off fuel would not adversely impact the type's pitch stability.

Rumpler continued bomber development with the 5A15. Based on the 4A15, the 5A15 used the same basic structure and engines but had a number of refinements. The wing span was increased slightly, the shape of the vertical tail surfaces was revised, and a single gravity tank under the top wing replaced the two separate gravity tanks over the engines. The gunner in the front cockpit was given a windshield and the rear crew member now had a flexible gun.

The 5A15 first flew on September 4, 1915 and demonstrated acceptable performance, although flight testing revealed a number of improvements were needed. The test pilot, probably Friedrich Budig, then chief test pilot at Rumpler, complained that the fuel system was too complicated and the gravity tank was the sole feed to the carburetors, creating a single point of failure. The aircraft was also too tail heavy and the pusher propellers were subject to damage from debris thrown up by the landing gear. Inadequate clearance (4 cm) between the fuselage and the propeller arc sometimes caused the propellers to hit the screen protecting the rear gunner from the propeller arc.

The 5A15 passed its acceptance flight on September 16, 1915, and *Idflieg* ordered it into limited production as the Rumpler G.I, only four aircraft being built. Photographs show the production aircraft had an enlarged rudder for better directional control after an engine failure. The very small number ordered indicates that the aircraft were primarily intended for operational evaluation. One aircraft, G.16/15, was delivered to *Kampgeschwader der Obersten Heeresleitung* 1 (*Kagohl* 1) for operational evaluation, where it was flown for a time by *Leutnant* Ray, while two others, G.17/15 and G.18/15, served briefly that year with *Brieftauben Abteilung Metz*, or *BAM*.

Below: Rumpler G.I 15/15 in flight.

Left: This photograph shows the Rumpler G.I engine installation in more detail. The under-wing gravity tank and main fuel tank for the starboard engine are clearly visible. The frontal radiator was large and complex.

Below & Bottom: Rumpler G.I 15/15, the first of four production machines. In this front quarter view the most visible difference between the Rumpler G.I and A415 is the windshield provided for the front gunner in addition to the gravity tank installed just below the upper wing. The front gunner's flexible machine gun is also visible.

Above: Rumpler G.I 15/15 is shown moments after lifting off during take-off. This view confirms that the enlarged rudder was retrofitted to this aircraft.

Bottom Facing Page: This rear view of a Rumpler G.I is probably 16/15, there is black and white measurement bar under the port wing. The engines were well streamlined but there were lots of rigging wires to create drag.

Below: This front quarter view of Rumpler G.I 16/15 shows the overall workman-like design and construction. The tents in the background indicate this is an operational aircraft at its unit.

Rumpler G.I Specifications

Engines:	2 x 150 hp Benz Bz.III	
Wing:	Span	19.3 m
	Area	78.68 m^2
General:	Length	11.8 m
	Height	4.0 m
	Empty Weight	1,998 kg
	Loaded Weight	2,938 kg
Maximum Speed:		145 kmh
Climb:	800m	7 min
	2000m	21 min
	4000m	120 min
Service Ceiling:		4000 m
Range:		600 km
Armament: 1 flexible MG & 150–200 kg of bombs		

Above & Below: Rumpler G.I 16/15, the second of four production machines. The most noticeable difference between this aircraft and the first G.I is the enlarged rudder with large aerodynamic balance for improved directional control with one engine out. The larger rudder was retrofitted to G.I 15/15. The engine cowlings were well streamlined. A black and white measurement bar is included in two photographs to enable measurements to be scaled from the photos.

Above: A Rumpler G.I on the Eastern Front with its lower starboard engine cowling removed for maintenance.

Rumpler G.I 16/15.

Rumpler G.II

In November 1915 *Idflieg* ordered 24 improved aircraft as the Rumpler G.II. The most important change was the use of more powerful, 220 hp Benz Bz.IV engines for improved speed, climb, and ceiling. The G.I airframe was used with minor modifications, but flight testing soon revealed the need for additional modifications for improved flying qualities. The prototype G.II, Rumpler internal designation 5A16, was tail heavy and insufficiently stable. Testing revealed the need for larger propellers to efficiently absorb the greater engine power, so propellers of 3.1 m diameter were fitted and the fuselage was narrowed slightly to provide sufficient clearance.

Rumpler chief test pilot Friedrich Budig determined that modifications to the lower wing near the propellers were needed, and on May 27, 1916 large cut-outs were made in the lower wing in front of the propellers on the second production aircraft. This modification greatly reduced tail heaviness as determined by Budig during a successful test flight on May 30.

Rumpler G.II 117/15, the last aircraft of the first production series, was fitted with more powerful 260 hp Mercedes D.IVa engines and larger propellers of 3.17 m diameter. *Idflieg* accepted this modified aircraft and it went on to serve with *Kagohl* 2 on the Eastern Front in the summer of 1916. No other G.II

Rumpler G.II Specifications

Engines:	2 x 220 hp Benz Bz.IV	
Wing:	Span	19.3 m
General:	Length	11.8 m
	Height	4.0 m
	Empty Weight	1,990 kg
	Loaded Weight	2,990 kg
Maximum Speed:		164 kmh
Armament: 2 flexible MGs & bombs		

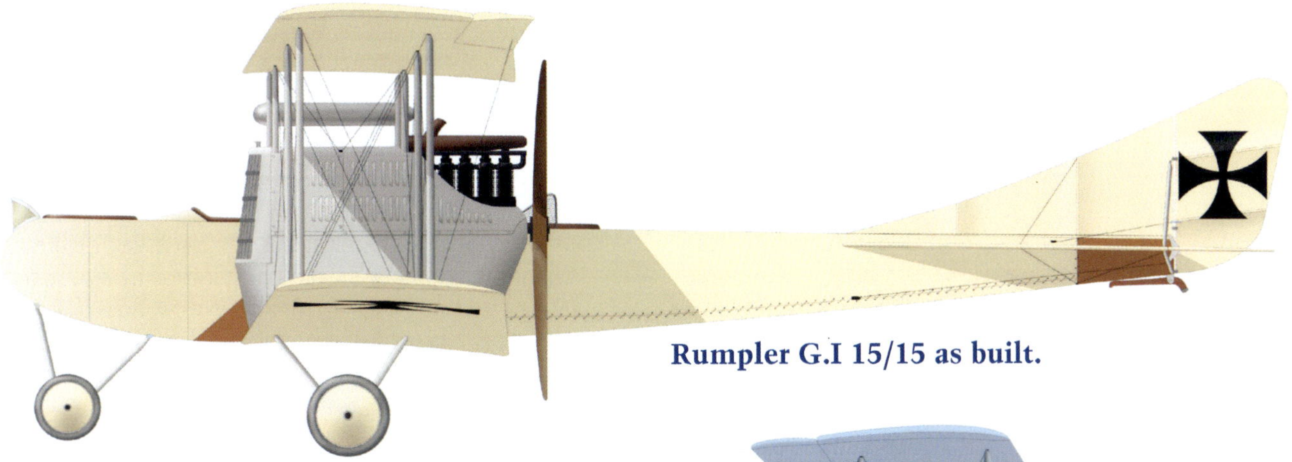

Rumpler G.I 15/15 as built.

Rumpler G.II G.109/15 assigned to *Kagohl* 2, summer of 1916. The black circles were identification markings painted at the unit level.

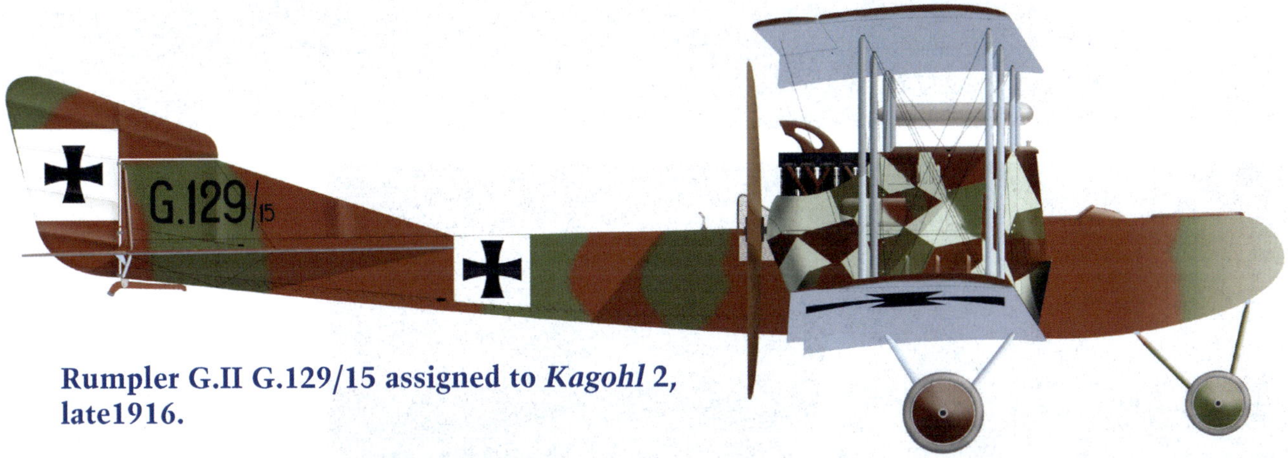

Rumpler G.II G.129/15 assigned to *Kagohl* 2, late1916.

Rumpler G.III G311/16 assigned to *Kasta* 9, summer of 1917. The black circles with white star and black/white wheel covers were unit markings.

Left: The Rumpler G.II was developed from the Rumpler G.I by replacing the 150 hp Benz Bz.III engines of the G.I with more powerful 220 hp Benz Bz.IV engines. This early G.II, possibly the first G.II airframe, is virtually indistinguishable from the G.I.

Left: The more powerful Rumpler G.II was much more difficult to fly than the G.I On 27 May 1916 large cutouts were made in each lower wing beneath the propellers on Rumpler G.II 107/15, the second G.II, as shown here. On 30 May Rumpler test pilot Friedrich Budig successfully flew this aircraft, and all subsequent G.II aircraft featured this cutout.

Below: The Rumpler G.II was developed from the Rumpler G.I by upgrading the engines. This G.II, 117/15, the last aircraft of the first production batch of G.II bombers, is virtually indistinguishable from the G.I other than the barely visible cutout of the lower wing trailing edge. This is also the one G.II modified to have 260 h.p. Mercedes D.IVa engines, but the photo is not clear enough to show which engine type is fitted.

aircraft used this engine, but it was a harbinger of things to come in the Rumpler G.III.

The Rumpler G.II saw much of its service with *Kagohl* 2. Originally formed from the *Brieftauben Abteilung Metz*, or *BAM*, in December 1915, *Kagohl* 2 had six *Kampfstaffeln*, *Kastas* 7–12. Five of these had single-engine C-type aircraft, but one was equipped with the Rumpler G.II and AEG G.III. *Kagohl* 2 was transferred to the Eastern Front in July 1916 and was eventually stationed at Lasnaja aerodrome. *Kagohl* 1, originally formed from the *Brieftaubenabteilung Ostend*, or *BAO*, was also transferred to the Eastern Front in June 1916, and had some Rumpler G.II aircraft along with other G-types. In September 1916 *Kagohl* 1 was transferred to northern Bulgaria where it flew a number of bombing missions against Romanian troops. *Kagohl* 2 was transferred back to the Western Front in October 1916, followed by *Kagohl* 1 in May 1917.

The Rumpler G.II bombers were better armed than the C-types and on the Eastern Front were used as long-range reconnaissance aircraft and as escorts for the smaller C-types on bombing raids.

Above: The most visible differences between Rumpler G.II, 117/15 and the earlier Rumpler G.I are the cutouts in the lower wing trailing edge and the white backgrounds for the iron cross insignia. The screens to protect the rear gunner from the propellers are also visible.

Right: Rumpler G.II G.107/15 is shown in flight over the Eastern Front while serving with *KG*2. The light overall finish still contrasts somewhat with the white backgrounds for the iron cross insignia. Both guns are visible as is the slight sweep-back of the wings.

Above: In contrast to earlier Rumpler bombers with overall light finish, Rumpler G.II G.129/15, the next to last production G.II, carries an interesting camouflage scheme, with the camouflage on the engine nacelles painted a different pattern than the fuselage. Despite the distinctive camouflage, only standard factory markings and insignia are visible. Because the factory serial number is clearly visible, the camouflage on wings and fuselage was probably painted at the factory. The straps for securing the bombs under the fuselage are clearly visible, as is the lower wing trailing edge cutout. The summer scene was probably taken at KG2 on the Eastern Front.

Below: This close-up view of Rumpler G.II G.129/15 shows its interesting camouflage scheme and engine nacelles in more detail. The camouflage appears to be sprayed on the fuselage and brush-painted on the nacelles, indicating the nacelle camouflage may have been applied at the unit. The screens to protect the gunner from the propellers are clearly shown. Although not clear, there appears to be segmented camouflage on the bottom of the upper wing.

Above: Rumpler G.II G.109/15 of *KG*2 damaged at Lasnaja, one of two fields *KG*2 is known to have used on the Eastern Front, where *KG*2 served from mid-July to October 1916. The black circle marking is distinctive. The radiators are very different than those fitted to most other Rumpler G.II bombers, but the reason for this modification is unknown.

Below: Then *Unteroffizier* Gustav Seitz poses in front of Rumpler G.II G.110/15 of *Kampfstaffel* 9 of *KG*2 at Kowel after receiving the Iron Cross 2nd Class. Again the radiators are distinctly different than those fitted to most other Rumpler G.II bombers.

Above: Rumpler G.II G.122/15 ready for a bombing mission; the bombs are already loaded under the fuselage. Only standard factory markings and insignia are visible. The snowy scene was probably at *KG*2 on the Eastern Front.

Above: An unknown Rumpler G is shown in the Rumpler factory. In contrast to all available known G.I photographs, this aircraft is in overall dark finish with light engine nacelles typical of the G.II, making one think this was a production G.II. There appears to be a lower wing trailing edge cutout characteristic of the G.II series but the angle and quality of the photo preclude positive verification.

Left: A Rumpler G.II in service with *BAM* (*Brieftauben Abteilung Metz*). Standing in the pilot's cockpit is *Hptm.* Fritz Prestien; the officer manning the gun is an observer named von Hachenburg. At the far right is *Lt.* Franz Ray, who would go on to score 17 confirmed victories as a fighter pilot. *Hptm.* Paul Backhaus is 3rd from right.

Below: The remains of Rumpler G.II G.130/15, the last production G.II, are shown here in Russian hands. Since no G.II bombers were downed by the Russians, it is a mystery how they retrieved this dismantled aircraft. G.130/15 appears to have a similar camouflage scheme to G.129/15, further indicating application at the factory.

Above: Rumpler G.II in flight is identified by its distinctive cutouts on the trailing edge of the lower wing that significantly reduced the design's tail heaviness in flight.

RUMPLER

Rumpler G.III

The Rumpler G.II had proven itself moderately successful and *Idflieg* placed an order for the proposed Rumpler G.III in September 1916. Retaining the basic size and configuration of the earlier G.II, the G.III was a new design. Although wingspan was the same as the earlier types, wing area was reduced by a smaller lower wing. The engine nacelles were more streamlined and more powerful 260 hp Mercedes D.IVa engines were fitted in place of the 220 hp Benz engines of the G.II. To reduce the danger of fire the fuel tanks were now mounted in the fuselage instead of in the nacelles. Unfortunately, the weight of the new aircraft increased substantially due to increases in the empty weight and also the designed payload.

Flight testing of the new prototype, factory designation 6G2, began in December 1916 and quickly revealed unsafe flying qualities. Like the earlier designs, the G.III prototype was tail heavy in flight. To solve this problem the horizontal stabilizer was raised and the fuselage near the plane of the propellers was modified to improve airflow, and these changes greatly alleviated the tail-heaviness. Test pilot Friedrich Budig stated the revised G.III was more stable and had better flying qualities than the G.II. However, adding the rear gunner's gun and ammunition again caused tail-heaviness, and the upper wings had to be moved rearward to restore pitch stability and trim.

Rumpler G.III Specifications

Engines:	2 x 260 hp Mercedes D.IVa	
Wing:	Span	19.3 m
	Area	78.68 m^2
General:	Length	12.0 m
	Height	4.5 m
	Empty Weight	2,365 kg
	Loaded Weight	3,620 kg
Maximum Speed:		150 kmh
Climb:	3000m	22 min
Service Ceiling:		5000 m
Range:		660 km
Armament: 2 flexible MGs & 225 kg of bombs		

The G.III that arrived at the front in December 1916 was the aircraft assigned to *Kagohl* 2, which had used the earlier G.II. Some G.III bombers served into at least March 1918 with *Kampfstaffel* 9 of *Kagohl* 2.

The final Rumpler bomber prototype, factory designation 6G4, had increased wing span and tractor-mounted 4-bladed propellers. The first test flight on 18 February 1918 indicated marginal flight characteristics when the aircraft swung to the right on takeoff and the pilot, Budig, was barely able to correct it. With extensive modifications necessary to improve its poor flying qualities and higher priority programs underway, further bomber development was abandoned.

Below: This Rumpler G.III shows its camouflage to advantage. The upper surfaces appear to be sprayed in two colors while the rudder is painted a single dark color and undersurfaces, struts, nacelles, and wheel covers appear to be in a single light shade. Propeller spinners are fitted, showing attention to streamlining details.

Above: This photograph of Rumpler G.III G.311/16 of *Bogohl* 2 is especially interesting because, unlike most G.III photos, it shows unit markings on the nose and the wheel covers are painted half dark and half light; black and white? The aircraft is loaded for a bombing mission with P.u.W. bombs under the fuselage along with a square *Kastenbombe* for blast effect.

Left: This front view of a Rumpler G.III emphasizes its clean lines with great attention to streamlining, a Rumpler hallmark.

Right: The front gunner's cockpit of the Rumpler G.III appears much more compact than the rear gunner's position despite housing a lot of equipment.

Above: Rumpler G.III photographed while serving with an operational unit.

Left: This front quarter view of a Rumpler G.III apparently at an operational unit shows its camouflage and clean lines to advantage. Interestingly, the rudder appears to be in two colors like the rest of the upper surfaces. The photo gives the impression of being of the same aircraft as the photo below, except the rudder appears to be in a single dark color. Are these two different aircraft or is this a trick of the light?

Right: This rear quarter view of a Rumpler G.III apparently at an operational unit shows that the propeller spinners were not fitted. The rear gunner's gun is fitted and his cockpit appears very roomy.

Above: Rumpler G.III G.304/16, the fifth G.III aircraft, is shown here on a rainy day at the Rumpler factory. The two-color factory camouflage with lighter engine nacelles and undersurfaces is clearly shown.

Below: This appears to be another photograph of Rumpler G.III G.304/16 in front of the Rumpler factory, but that cannot be confirmed because the serial number is not legible. This view shows the cleanly-streamlined engine nacelles to advantage. The conical propeller spinners are not fitted in this view. Like the G.II, the G.III featured screens to protect the rear gunner from the propellers. The rudder and horn balance appear larger than that of the Rumpler G.II was would be expected for control of a more powerful aircraft in engine-out situations.

Right: Rumpler G.III, perhaps G.304/16? The photo looks like it could be from the same series as the other two on this page.

Left: This over-exposed view of a Rumpler G.III in flight is still interesting. The rudder and horn balance appear somewhat larger than that of the G.II and the lines are cleaner than the competing Gotha bombers. The nose landing gear used in all Rumpler bombers helped reduce the likelihood of a landing accident.

Below: This front view of a Rumpler G.III in a hangar shows a diamond insignia on the nose, which was a unit or personal marking.

Above: This G.III is shown under construction at the factory. Although the photo quality is poor, the revised engine nacelles that now were raised above the wing like the Gotha G.V are clearly visible.

Right: This photo taken from the front gunner's cockpit shows Rumpler test pilot Friedrich Budig (at right) flying a G.III. Only the pilot had a windscreen.

Rumplers After WWI

Rumplers were to serve in a number of countries post-Armistice as both military and civil aircraft.

Belgium had five C.IVs on the civil register. O-BABO; O-BAHI; OBUIS; O-BOND; and 0-BORD.

Czechoslovakia operated Rubild Mb 7955/18 in 1921.

Denmark operated at least one civil Rumpler C.I.

Finland also used one C.VIII. This carried the serials F28, C74/18, 2B350 and finally 2C350. It apparently survived until February 1925. Finland also operated one 6B1 and one 6B2 floatplane. The 6B1 was serialled F20, D60-18, and 1C310. Obtained from Germany in 1918, it served with the Aviation Battalion at Santahamina in the summer of 1918, and was written off on 25 October 1919. The 6B2 was reportedly purchased from the Germans in Tallinn in 1918. Given the serial 5A1, it was in service from November 1918 to October 1926, being written off the following September with 35 flying hours recorded in Finnish service. This machine is preserved in the Finnish Air Force aviation museum.

The Latvian Aviation Park operated at least four Rumpler C.I biplanes.

Netherlands

Three Rumpler C.I biplanes were interned in the Netherlands.

1. C.Ia(Han) 968/17. Landed at Elburg on 223 August 1917. The pilot, *Lt*. Heinz Brandt, had become lost and made a successful landing, the undamaged aircraft being flown to Soesterberg. Given the LVA serial LA-49 (later R-409), it was not purchased.
2. C.I(Mark) 2560/17 was impressed as R-410 after landing at Zesgehuchten on 1 December 1917. Crashed on 5 February 1918, the engine was still of use and later was fitted to Albatros C.III AL-411.
3. C.I 8364/17 had two technicians as crew. They had become lost in fog and landed the unarmed aircraft. R-480 in LVA service. This aircraft is reported to have been written off in 1920.

The Netherlands had great difficulty in obtaining up to date military equipment during the years of WWI, particularly in that they needed to defend their neutrality by preventing incursions by aircraft of the combating powers. The *Luchtvaartagdeeling* (*LVA*) of the Army obtained most of its aircraft by interning those of the combatants that fell within the Netherlands. These included the Rumpler C.I reconnaissance biplanes referred to above. In 1917 the Germans were willing to sell more modern aircraft to the LVA and ten Fokker D.III fighters were obtained, the first arriving at Soesterberg early in October. The Germans were willing to make Rumpler C.VIII trainers available for what was considered of more importance to the war effort; they wanted 5,000 horses!

The LVA received 40 C.VIII trainers in total from April 1918; however, many were stored due to a lack of trained staff to assemble them. The Argus engines proved troublesome and were prone to breaking fuel feed pipes due to vibration and the engine would catch fire. The Dutch wrapped the pipes with rubber to eliminate the problem. After the war it was learned that the *Fliegertruppe* had overcome this problem by repositioning the Zenith carburetor and installing new fuel piping. The Rumpler was different from previous *LVA*-operated aircraft and used a wheel rather than a stick control. It was sensitive to the rudder. The machines arrived in three batches and by the time the second batch arrived only four of the initial eight C.VII trainers remained. The type gained such a reputation that only volunteers were trained on the type. With the arrival of Fokker aircraft the Rumplers were grounded in November 1919, but it was not until 1925 that the Rumplers together with other interned material were sold off. Three were turned into taxi trainers, and others were used for other training. As late as 1934 some fuselages were still in storage.[1]

Poland used a number of Rumpler types in its air service. A 1919 tabulation gives the following: C.I – 21; C.IV – 10; C.V – 1; C.VI – 1 and C.VII – 2.

Soviet Russia had at least one C.IV in service.

Turkey operated the Rumpler B.I before receiving 10 Rumpler C.I biplanes in the spring of 1917 and they played a part in the battles for Gaza. Turkey may have had one C.IV postwar but this is not confirmed.

A number of Rumpler C types were brought back to the USA after the war. Four Rumplers identified as 'C-4 biplanes' were offered for sale. At least one was used as a civil aircraft in Florida with Mabel Cody's Flying Circus with the logo "Coral Gables" under the lower wing advertising a housing estate.

At least 19 Rumpler C.I biplanes were on the German civil registry as well as two C.IV biplanes. Some were given limousine cabins that could accommodate two passengers. (Courtesy Colin Owers)

1. Gerdessen, FG. "Dutch Rumpler C.VIII", *Small Air Forces Observer*, Vol.34, No.4, April 2011, page 119.

Rumpler Aircraft in Retrospect

Rumpler was one of the most innovative and dynamic aircraft manufacturers in Germany. Although producing a variety of warplanes, Rumpler will always be famous for the exceptional Rumpler C.IV and the family of aircraft that used its airframe; these were the best German long-range, high-altitude reconnaissance airplanes and were almost immune from interception by Allied fighters while undertaking their missions. The Rumpler C.X just entering production at the Armistice looked ready to continue the example of the C.IV family.

In addition, Rumpler built a small number of early reconnaissance floatplanes for the Navy, and built good B-type two-seaters and the excellent C.I, which performed well in combat and was built and used as a trainer until the end of the war due to its robust airframe with good flying qualities.

The transitional C.III was a modest success at the front but was very useful in sorting out the high-performance airframe for the more powerful C.IV that quickly followed. The C.III and C.IV airframes were lightly built to achieve good climb and ceiling, resulting in a number of modifications to strengthen the airframe based on operational experience.

Rumpler's achievements in fighters were more unusual. Oddly, the most successful Rumpler fighter was the 6B1 floatplane fighter derived from the C.I airframe. Rumpler's prolonged attempt to deliver a high-altitude 'superplane' was frustrated by reaching too far toward extreme performance at the expense of practicality. Despite investment of immense engineering and development resources using innovative structures and aerodynamics, the final result was failure to bring a practical fighter into production and operational service. Possibly the Armistice came too soon for Rumpler, or perhaps the airplane would never have gelled into a practical front-line fighter. Certainly by late 1918 other companies were delivering aircraft of similar exceptional performance without the limitations of the Rumpler D.I, especially SSW with its D.VI parasol monoplane and Daimler with its L11 parasol monoplane. Both of these aircraft looked to be practical production fighters with performance comparable to the Rumpler and perhaps exceeding it, especially in speed and maneuverability.

Rumpler's bombers were moderately successful on operations and were produced in small numbers. Although a number were lost in accidents, they had the rare distinction of none being shot down by enemy action. Perhaps their biggest limitation was mediocre flying qualities, with tail-heaviness a problem that persisted throughout the series.

Given that Rumpler was producing the best high-altitude reconnaissance airplanes in Germany, a type essential to the German war effort, and was struggling with developing competitive bombers, something had to be done. That something was cancelling further Rumpler bomber development to enable Rumpler to focus on what it was most successful at doing; designing and building high-altitude reconnaissance airplanes.

Above: Rumpler *Taube* floatplane. The German navy purchased at least three Rumpler *Taube* floatplanes, Marine Numbers E1, E4, and E8, all powered by 100 hp Argus engines.

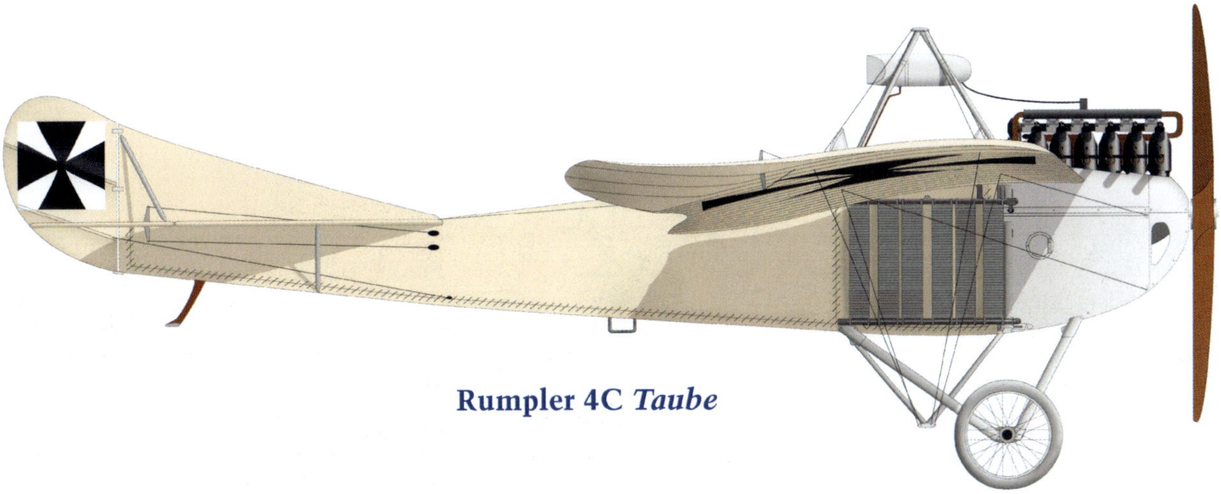

Rumpler 4C *Taube*

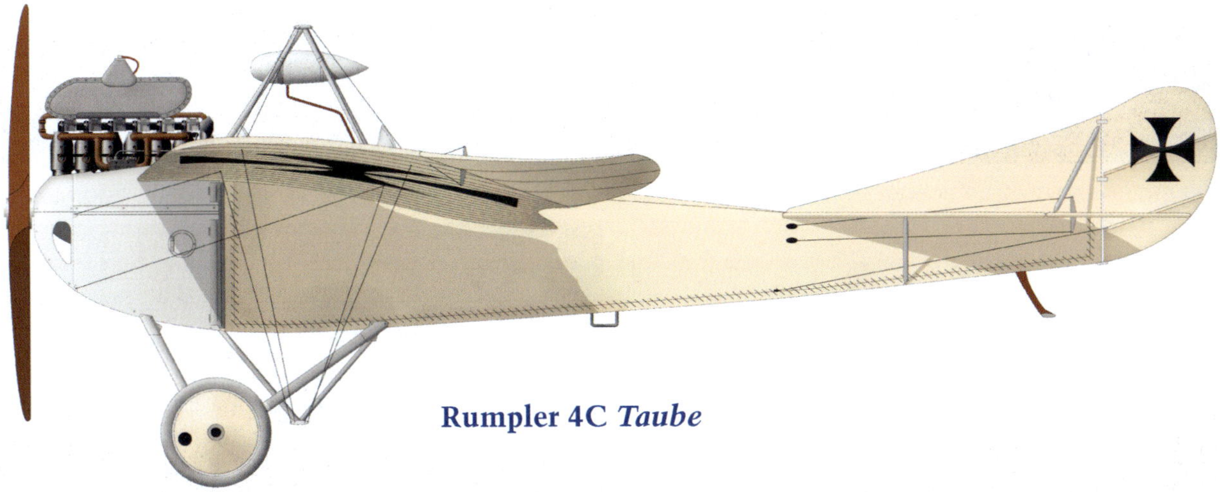

Rumpler 4C *Taube*

Rumpler C.III of the Wireless Operators Training School

Selected Bibliography

Books

Armin, Jörg, *Edmund Rumpler – Wegbereiter der industriellen Flugzeugfertigung*, Bernard & Graefe Verlag, 2004

Franks, N.L.R., Bailey, F.W, and Guest, R., *Above the Lines*, Grub Street, 1993.

Franks, N.L.R., Bailey, F.W, and Duiven, R., *The Jasta Pilots*, Grub Street, 1996.

Gray, Peter, and Thetford, Owen, *German Aircraft of the First World War*, second revised edition, New York: Doubleday & Company, Inc., 1970.

Green, William, and Swanborough, Gordon, *The Complete Book of Fighters*, Smithmark Publishers, 1994. Grosz, P.M., *LFG Roland C.II*, Albatros Productions, Ltd., 1995.

Grosz, P.M., *Rumpler C.I*, Albatros Productions, Ltd., 2000.

Grosz, P.M., *Rumpler C.IV*, Albatros Productions, Ltd., 1992.

Grosz, P.M., *Rumpler D.I*, Albatros Productions, Ltd., 1998.

Grosz, P.M., *The Taube at War*, Albatros Productions, Ltd., 2004.

Herris, J.W., *Development of German Warplanes in WWI*, Aeronaut Books, 2012.

Herris, J.W., *German Seaplane Fighters of WWI*, Aeronaut Books, 2012.

O'Connor, N.W., *Aviation Awards of Imperial Germany in World War I and the Men Who Earned Them*, Vols. 1–7, Foundation for Aviation WWI and Flying Machines Press, 1988–1999, Schiffer 2002.

Articles

Chapman, Peter, & Ansell, Richard, "The Rumpler G.I–III" *Over the Front* Vol.26 No.4, Winter 2011, p.328–349.

Grosz, Peter M., "Frontbestand" *WW1 Aero* No.107, Dec. 1985, p.60–66.

Grosz, Peter M., "Frontbestand" *WW1 Aero* No.108, Feb. 1986, p.66–69.

Internet

http://en.wikipedia.org/wiki/Edmund_Rumpler

http://en.wikipedia.org/wiki/Rumpler_Tropfenwagen

Selected Index

Above: Rumpler 6B1 #1046 in Allied colors.

Above: Rumpler G.II crash.

Rumpler C.I

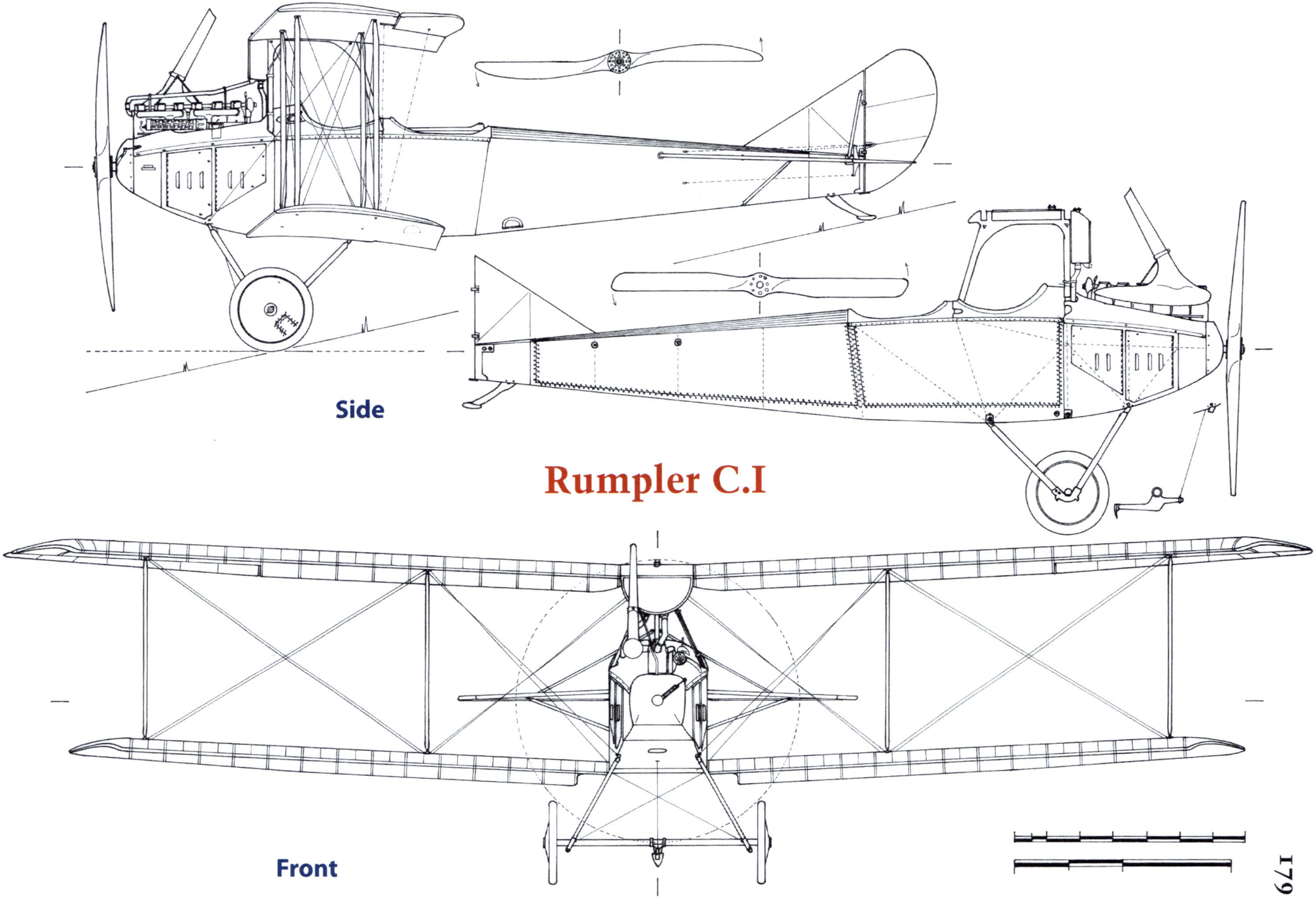

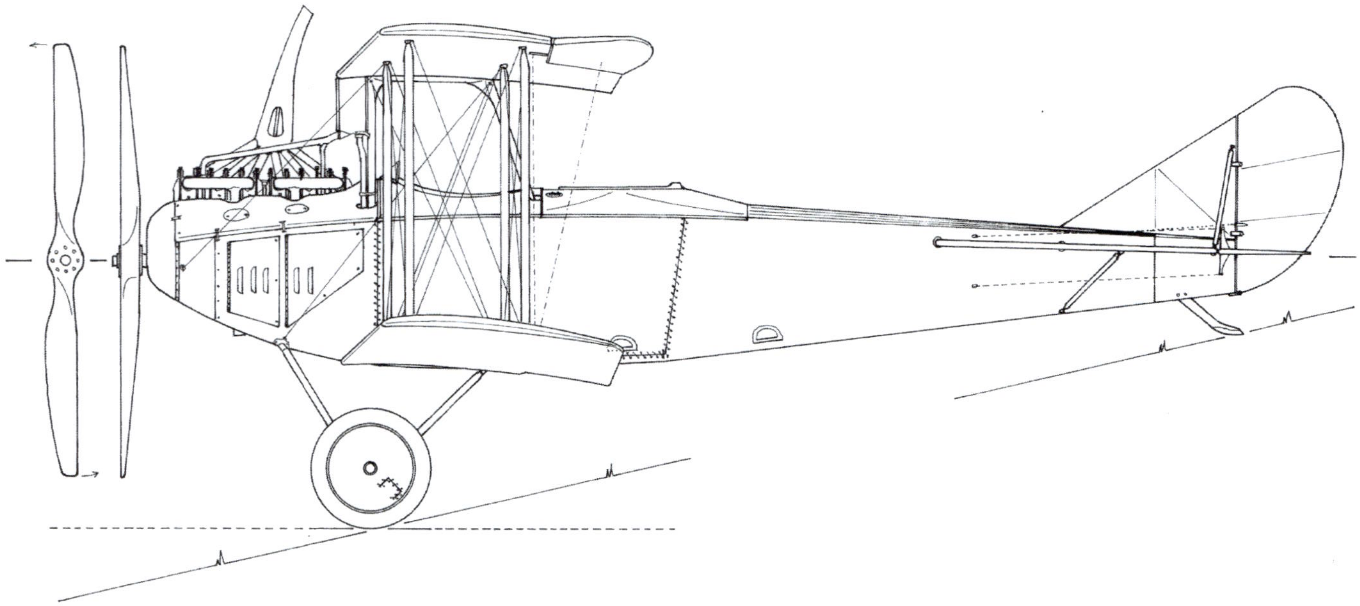

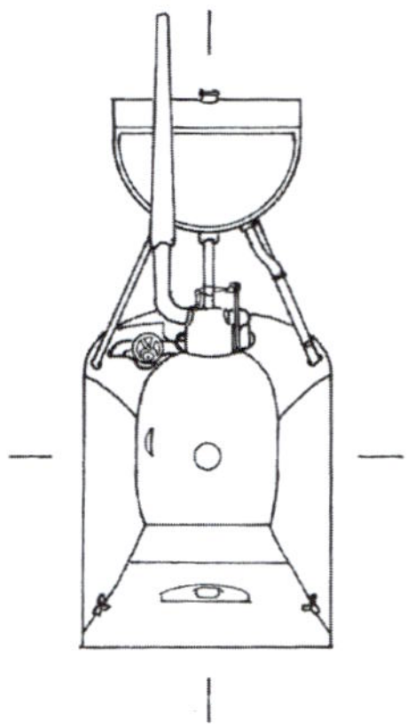

Front

Nose and fixed
gun details

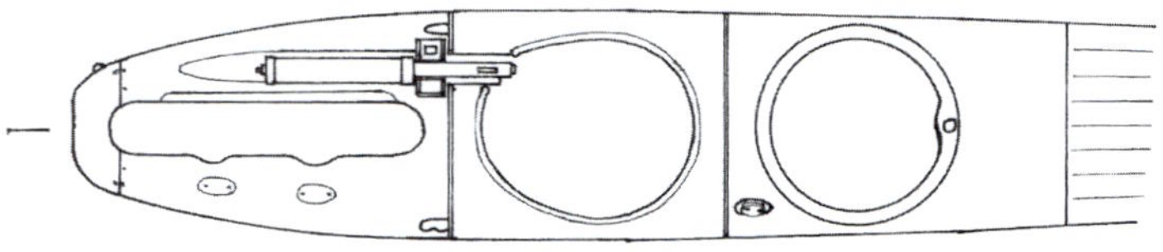

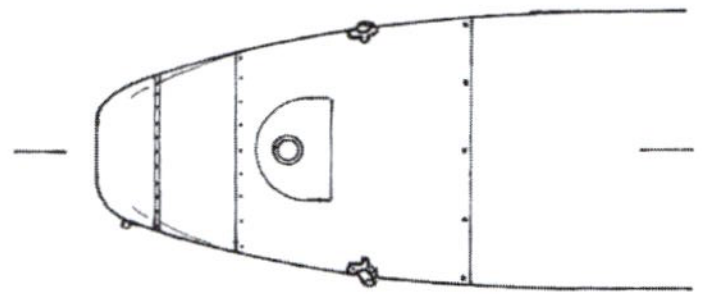

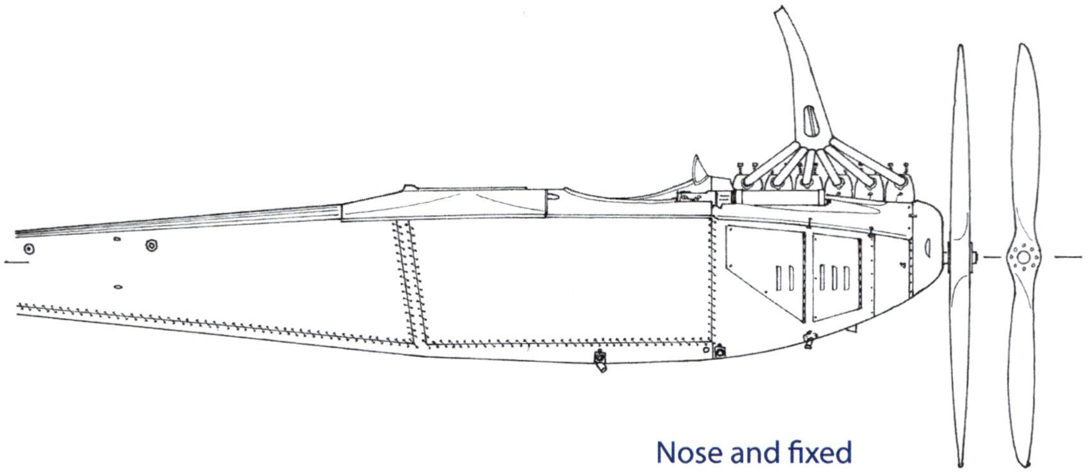

Nose and fixed
gun details

Rumpler C.Ia

Side

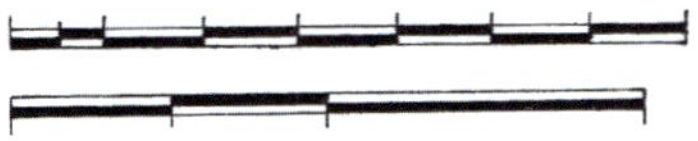

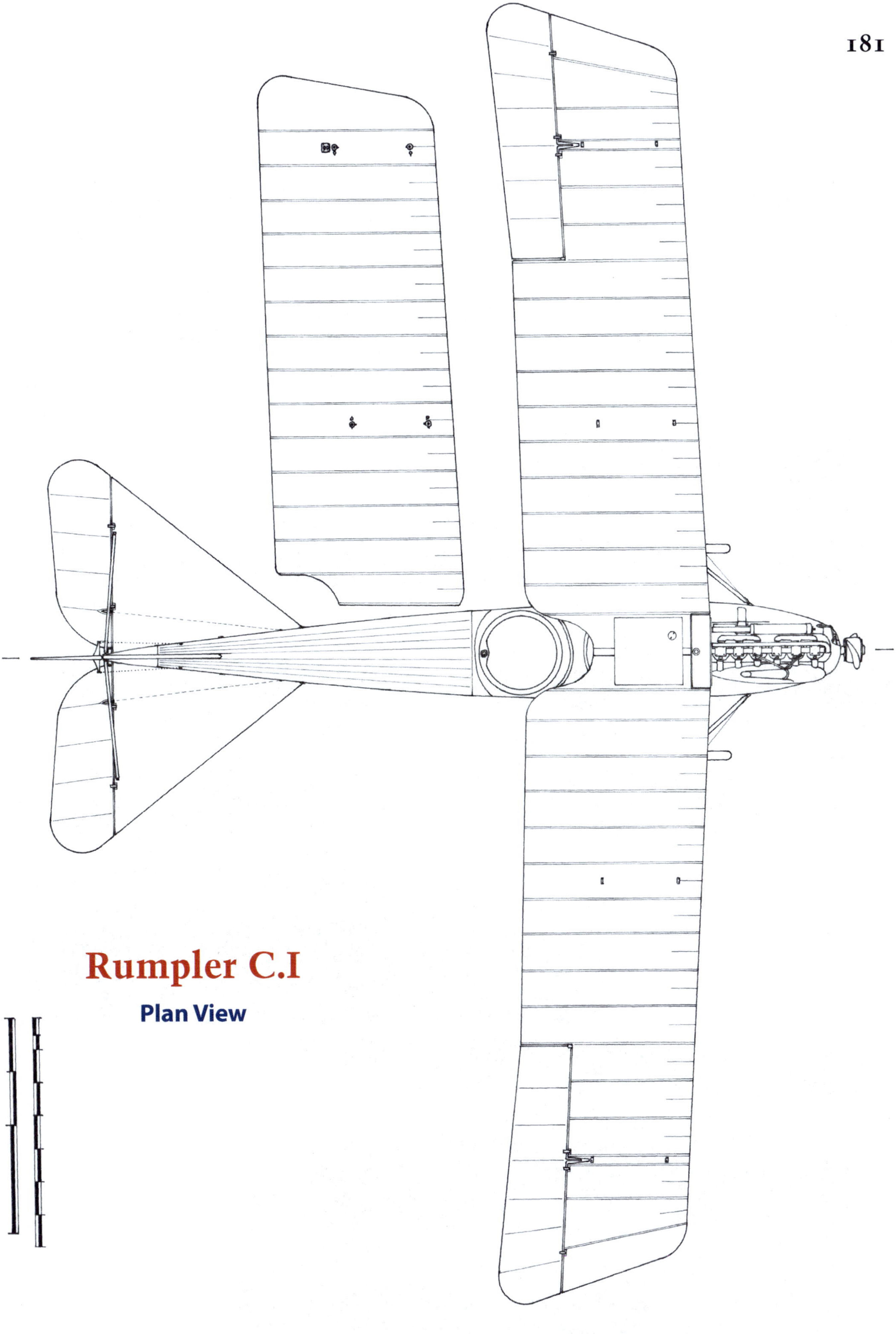

Rumpler C.I

Plan View

Rumpler C.I

Underside

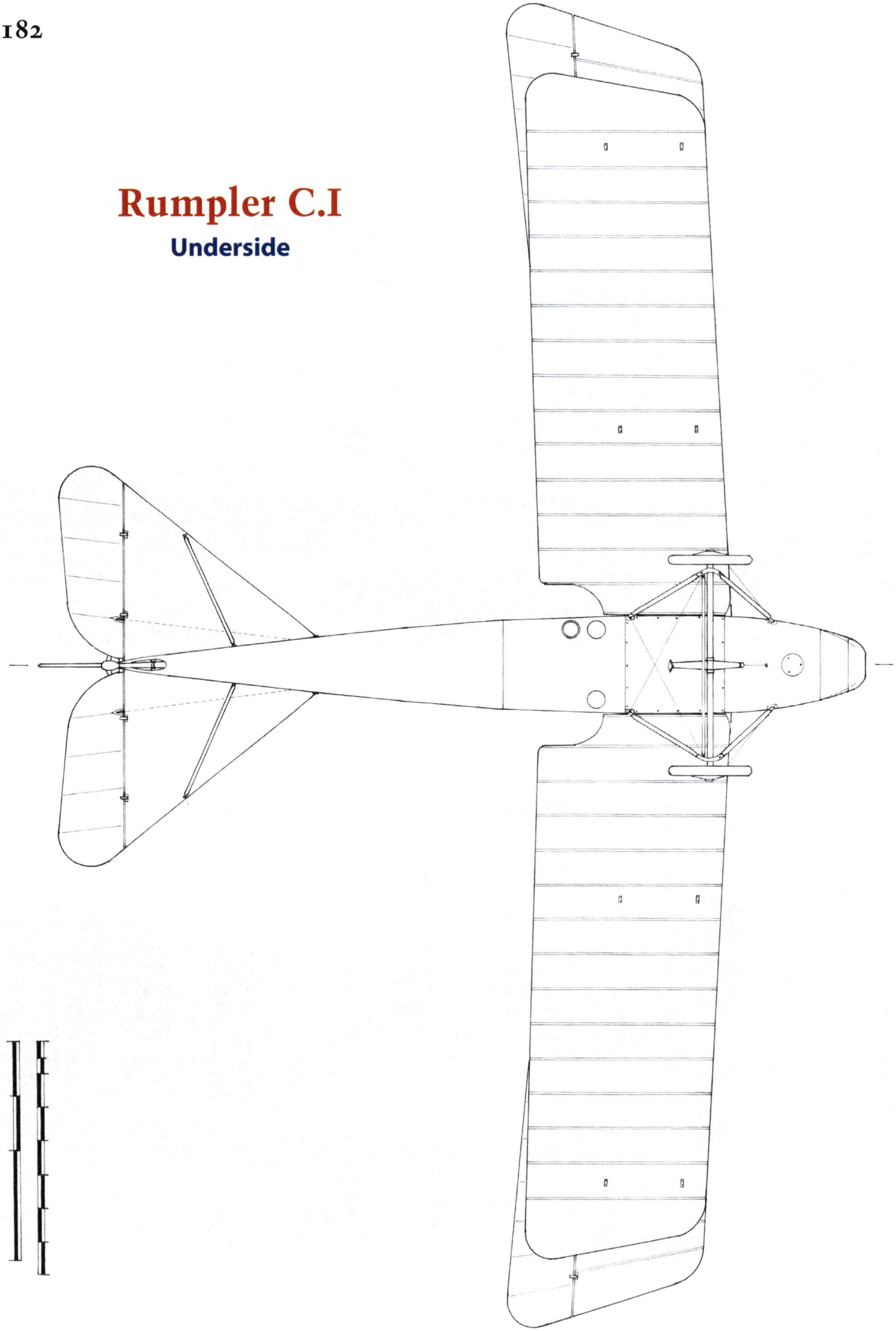

Rumpler C.III

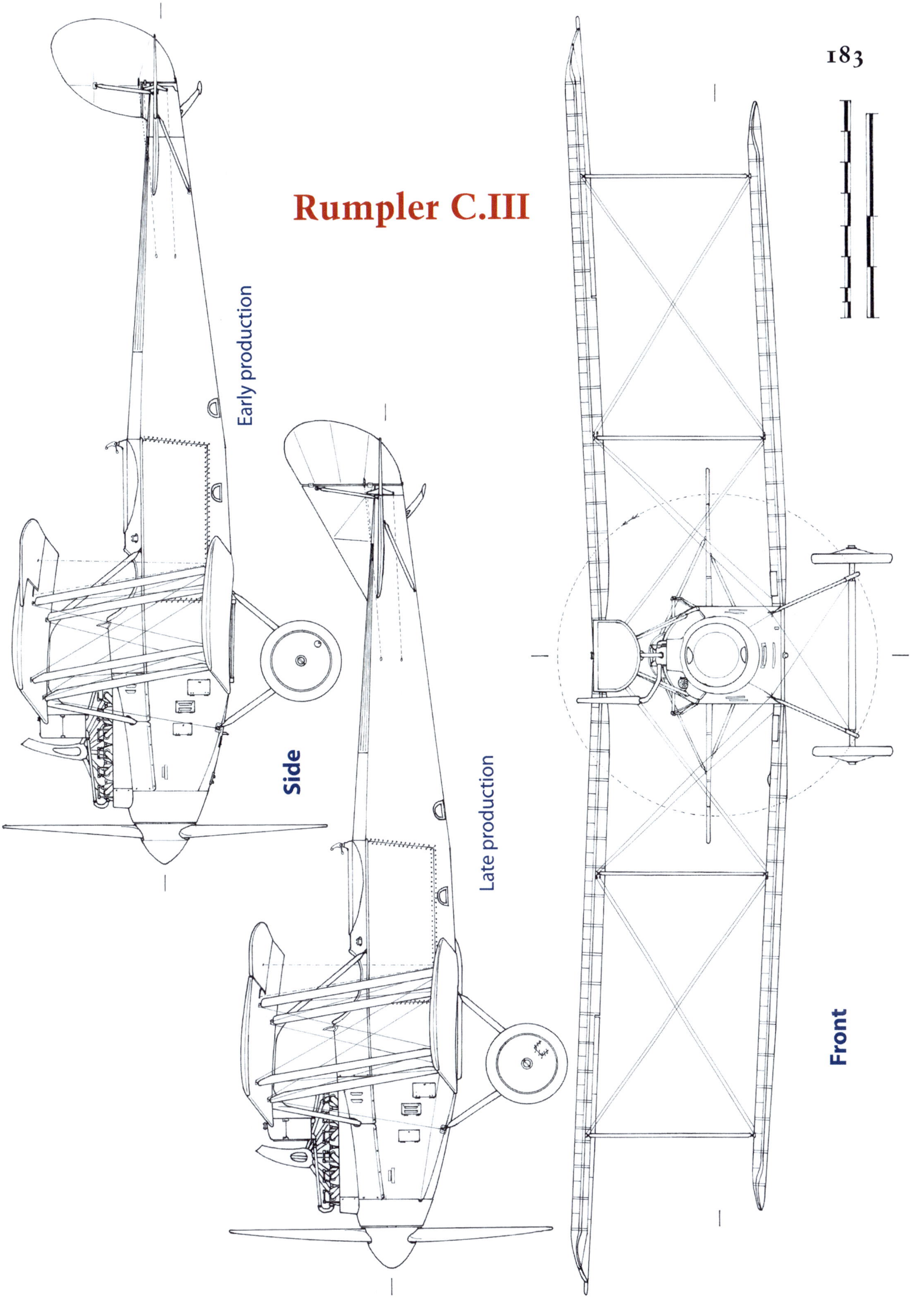

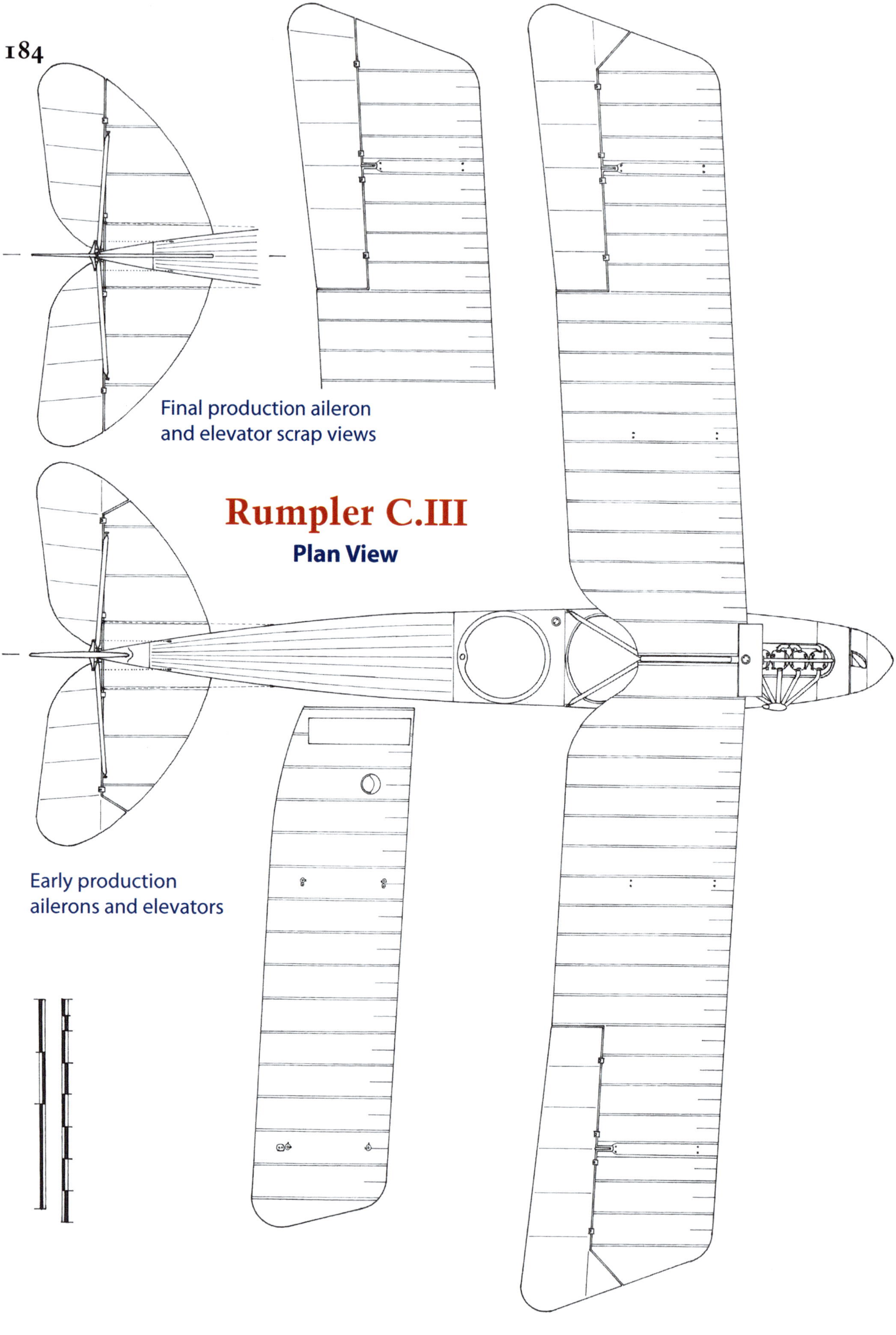
Final production aileron
and elevator scrap views
Rumpler C.III
Plan View
Early production
ailerons and elevators

Rumpler C.III

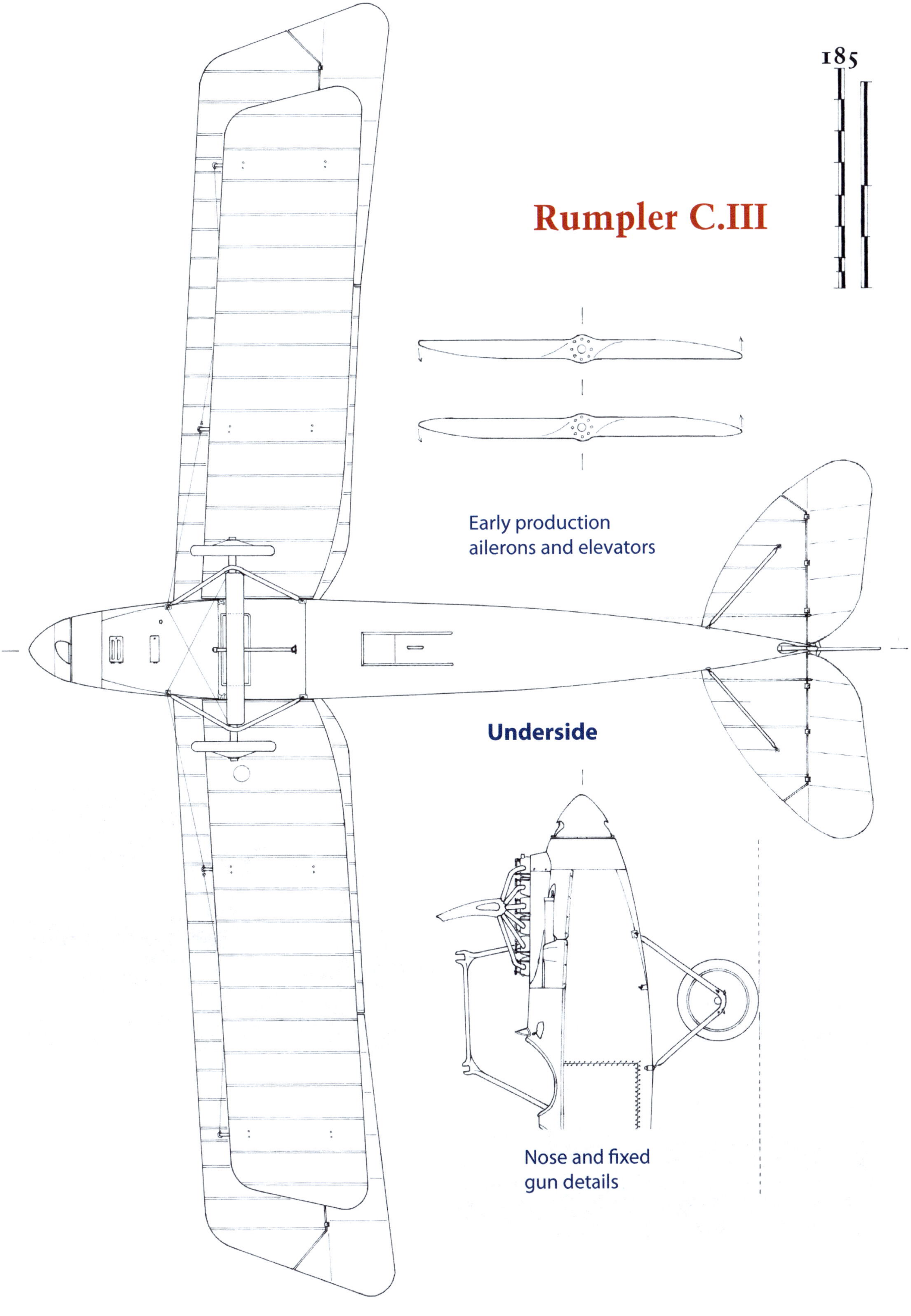

Rumpler C.IV

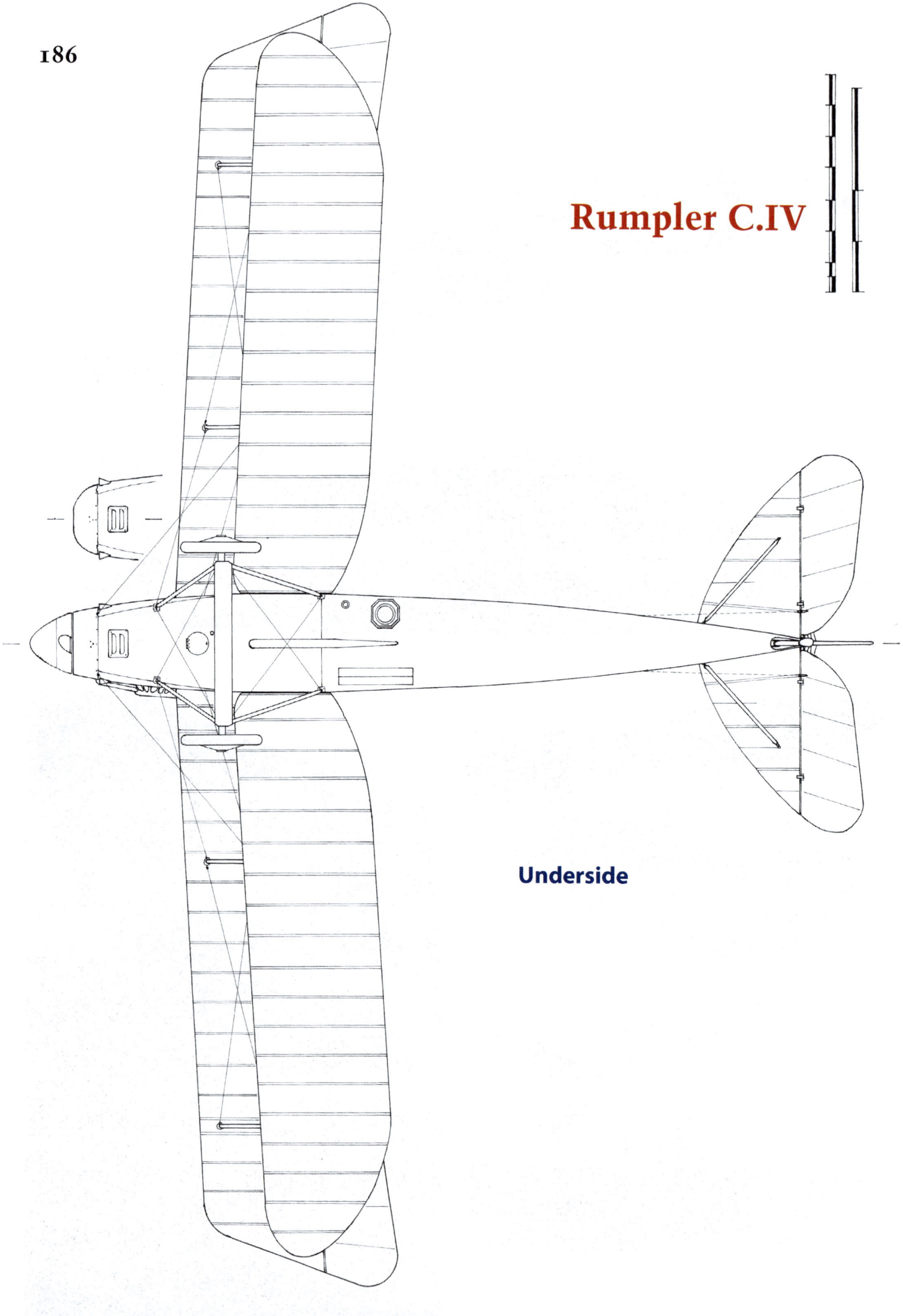

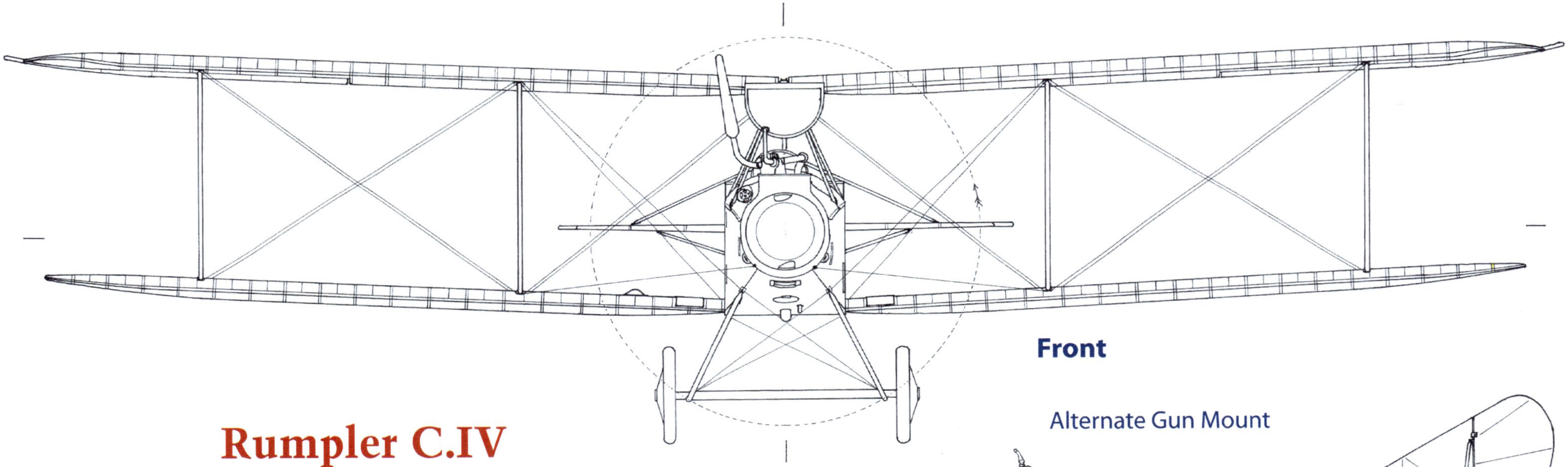

Front

Rumpler C.IV

Alternate Gun Mount

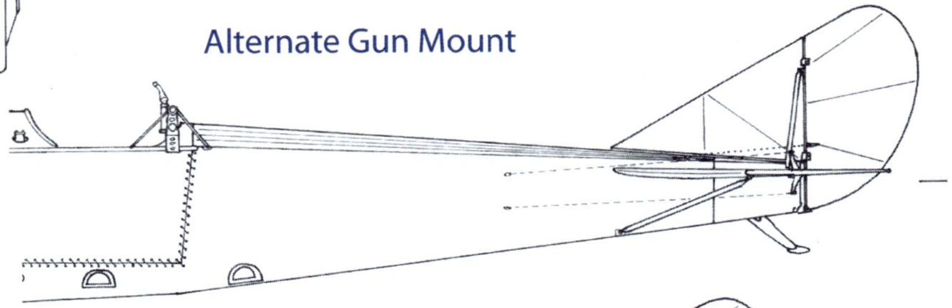

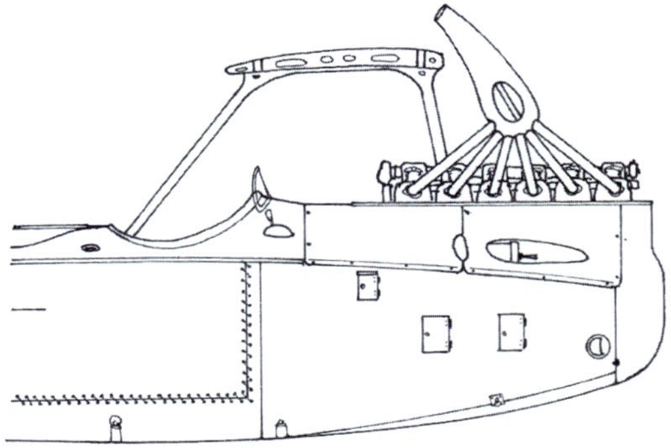

Nose and fixed gun details

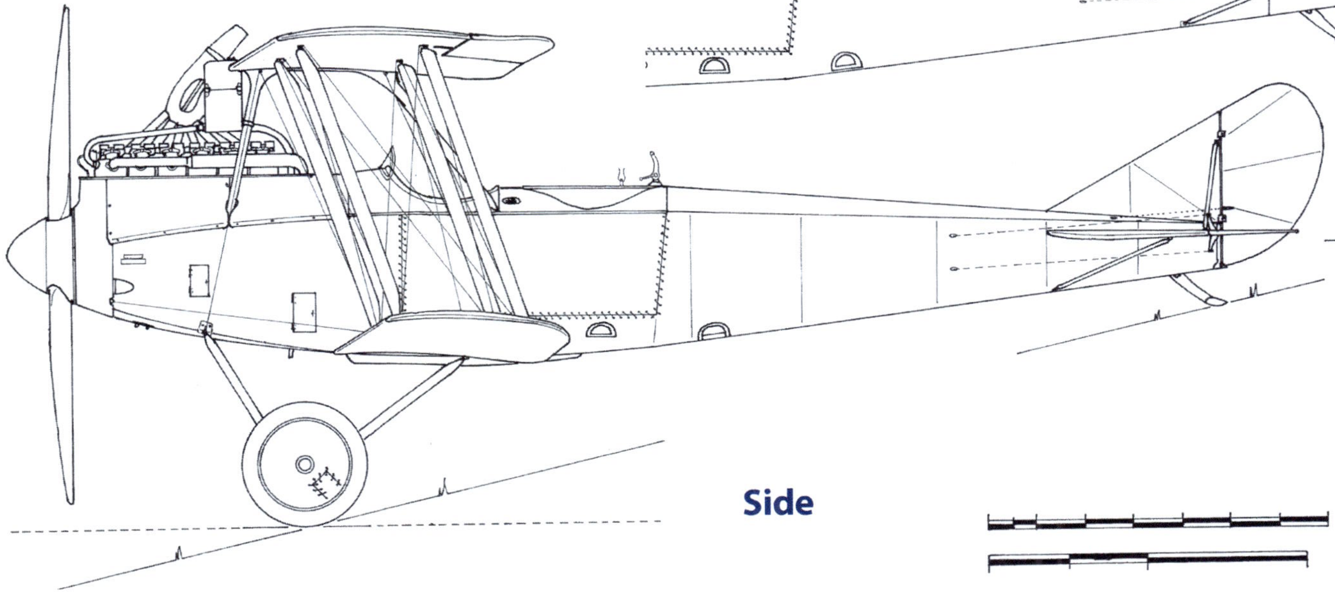

Side

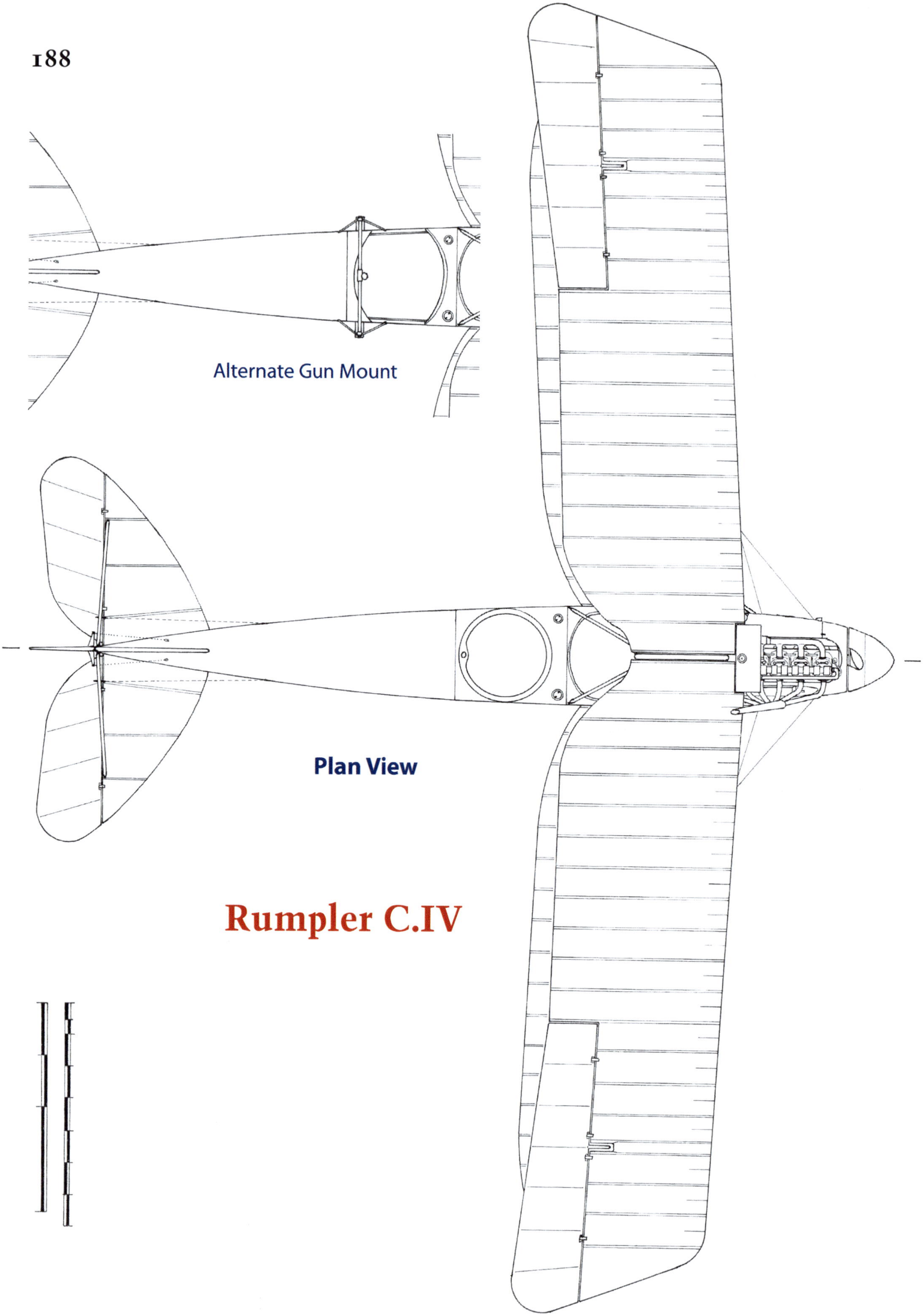

Alternate Gun Mount

Plan View

Rumpler C.IV

Rumpler C.X

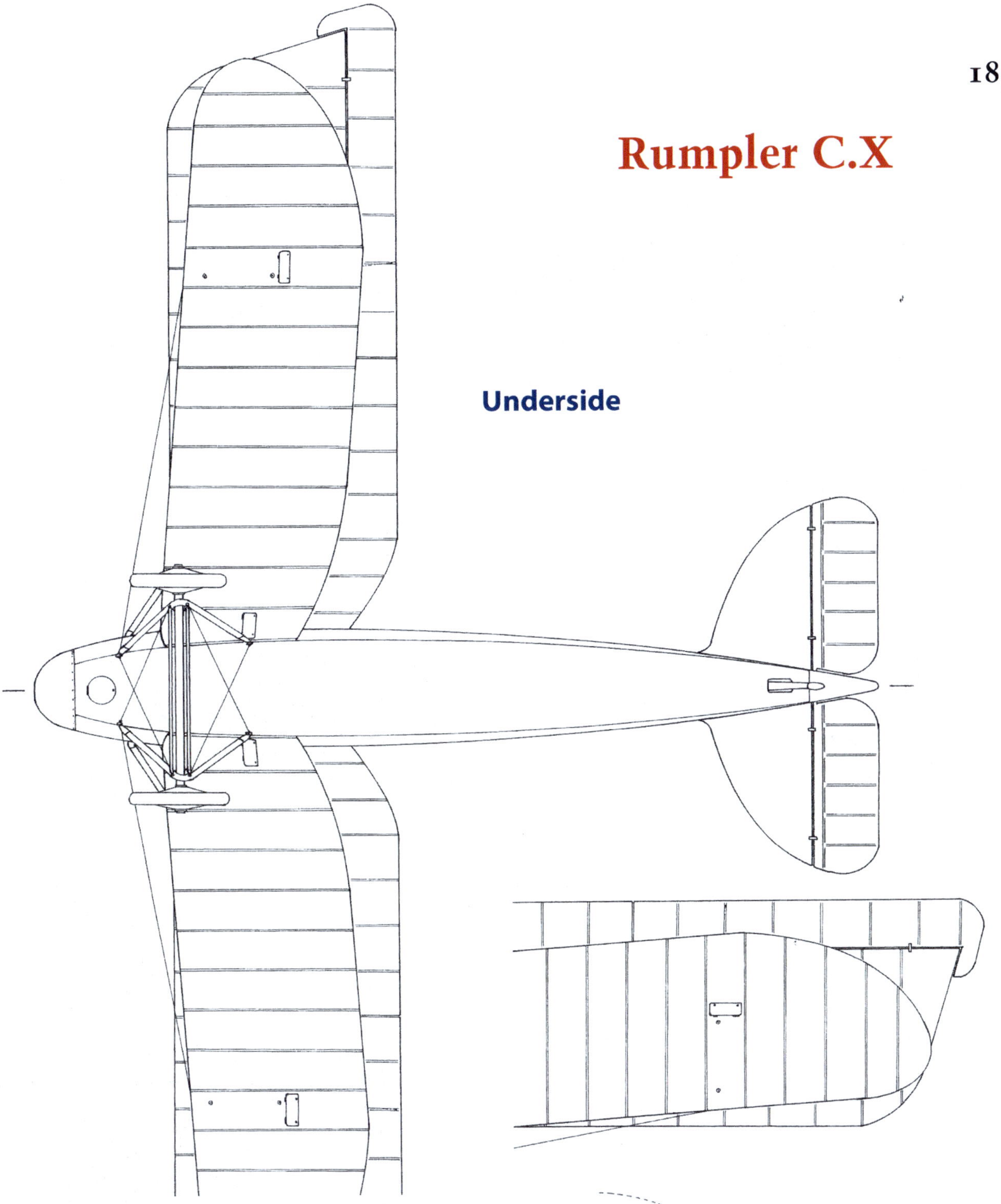

Underside

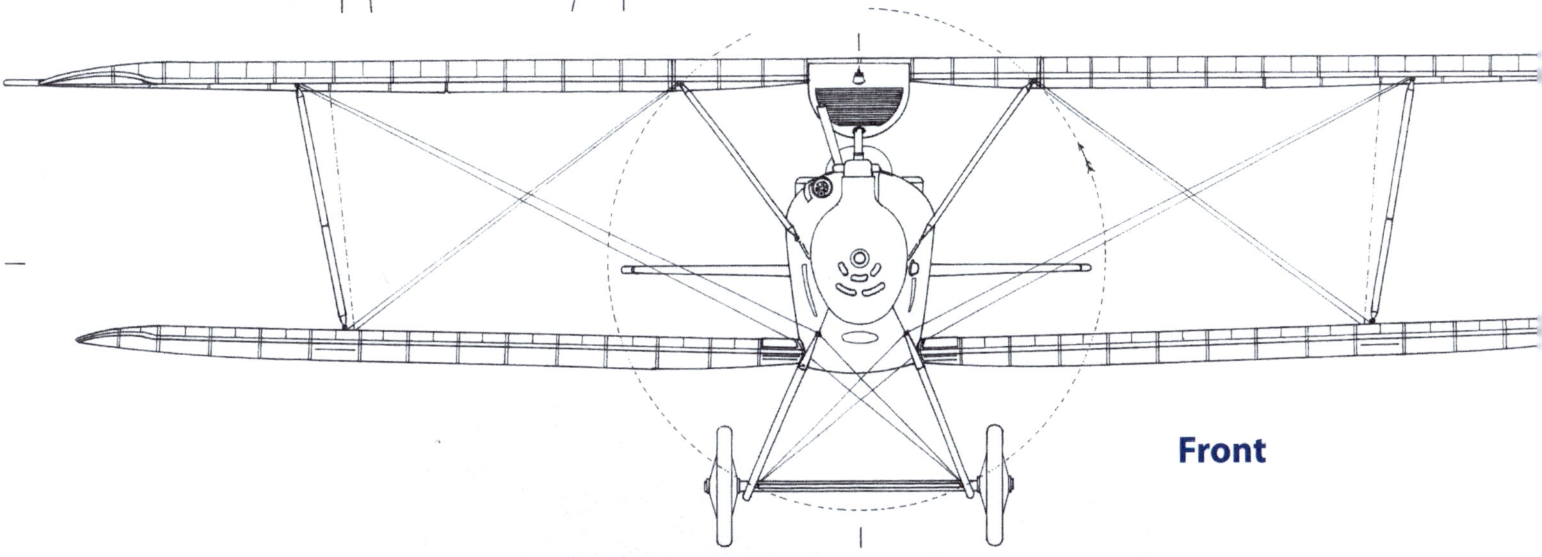

Front

Rumpler C.X

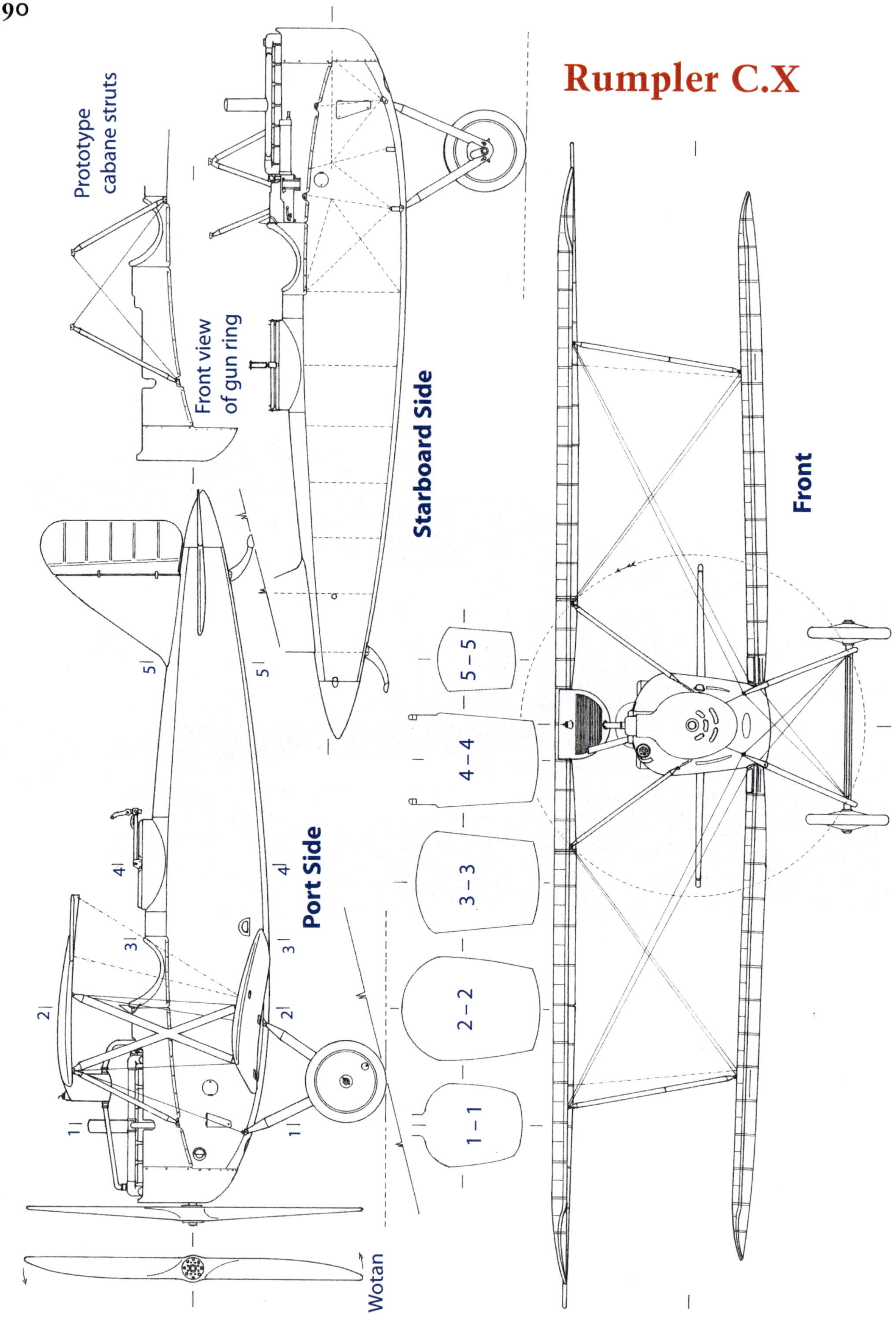
Prototype cabane struts
Front view of gun ring
Starboard Side
Port Side
Front
Wotan
1–1
2–2
3–3
4–4
5–5
1
2
3
4
5

Rumpler C.X

Plan View

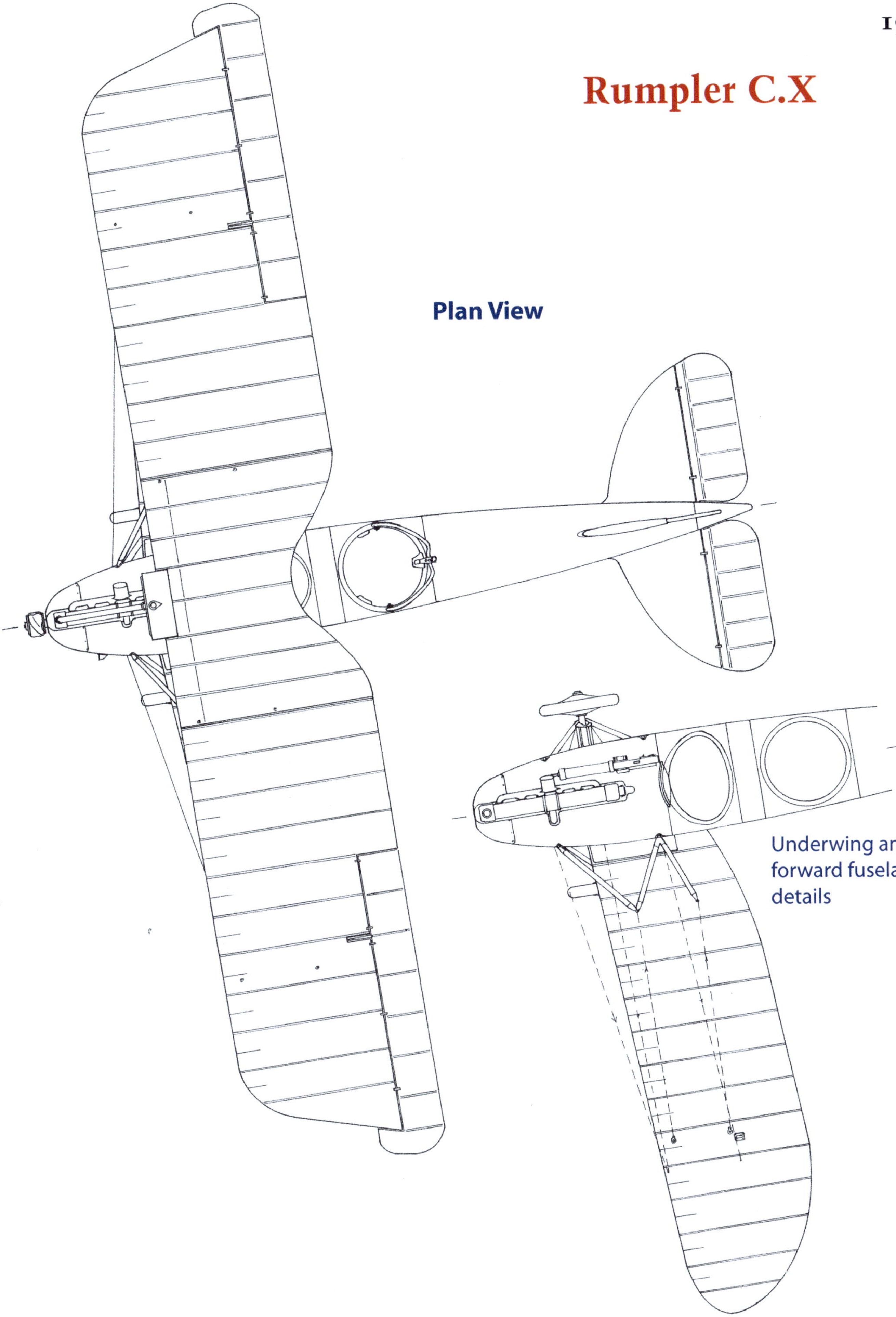

Underwing and forward fuselage details

Rumpler 6B1

Plan View

The three 6B1 prototypes, Marine Numbers 751, 787, and 788, used the straight-edged horizontal tailplane from the Rumpler C.I. All production 6B1 fighters used the curved tailplane from the Rumpler C.IV.

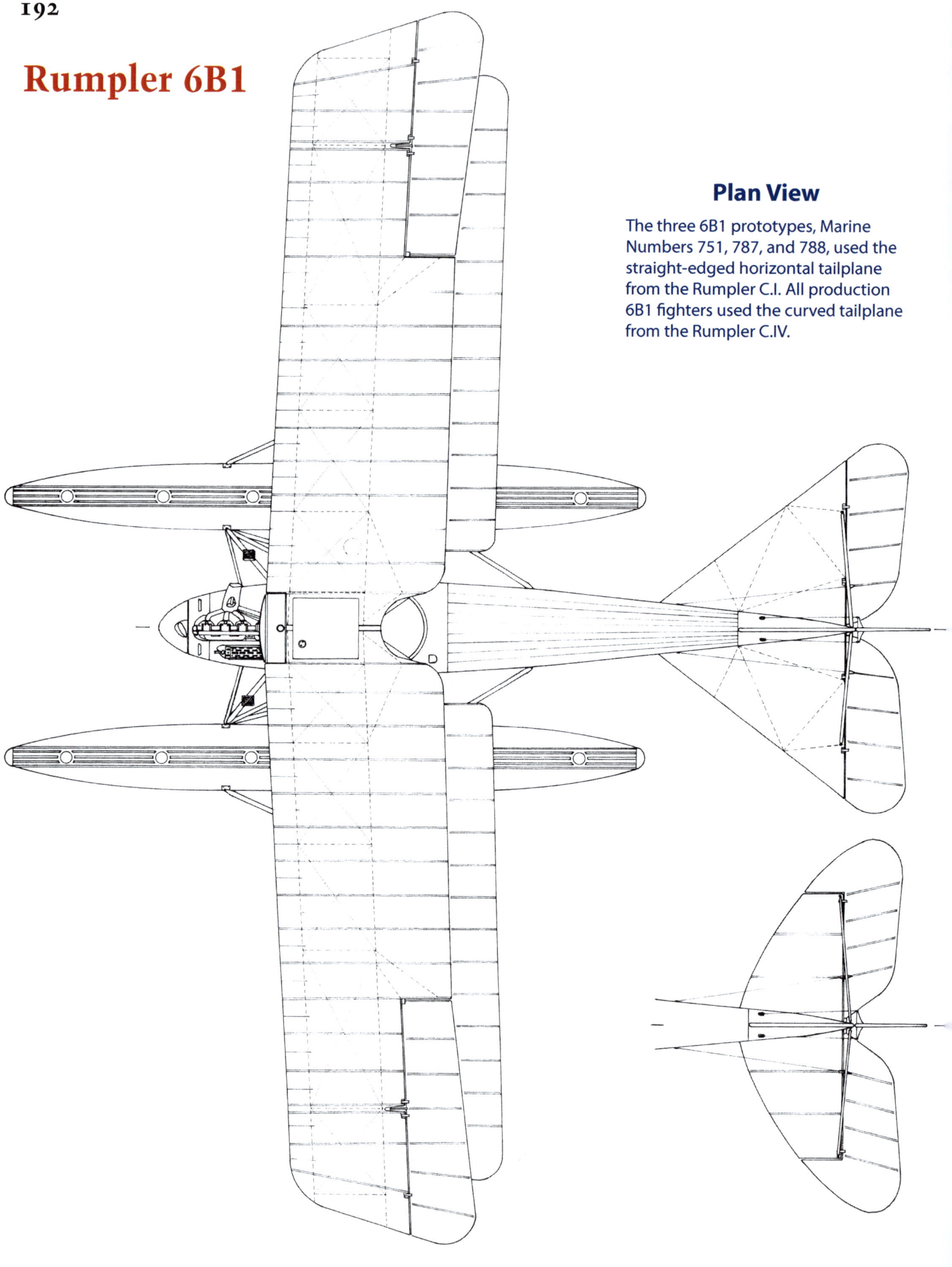

Rumpler 6B1

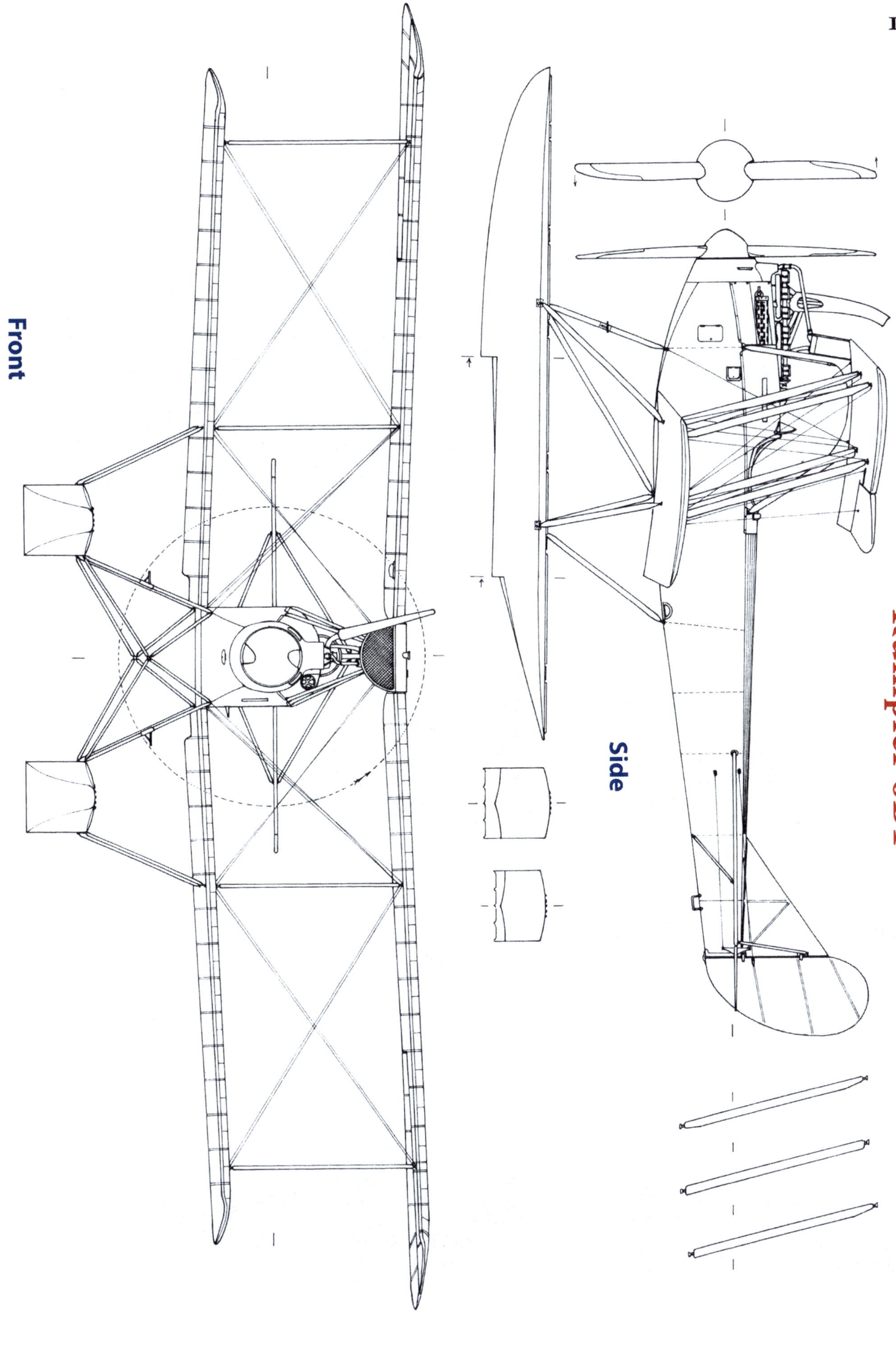

Rumpler 6B1

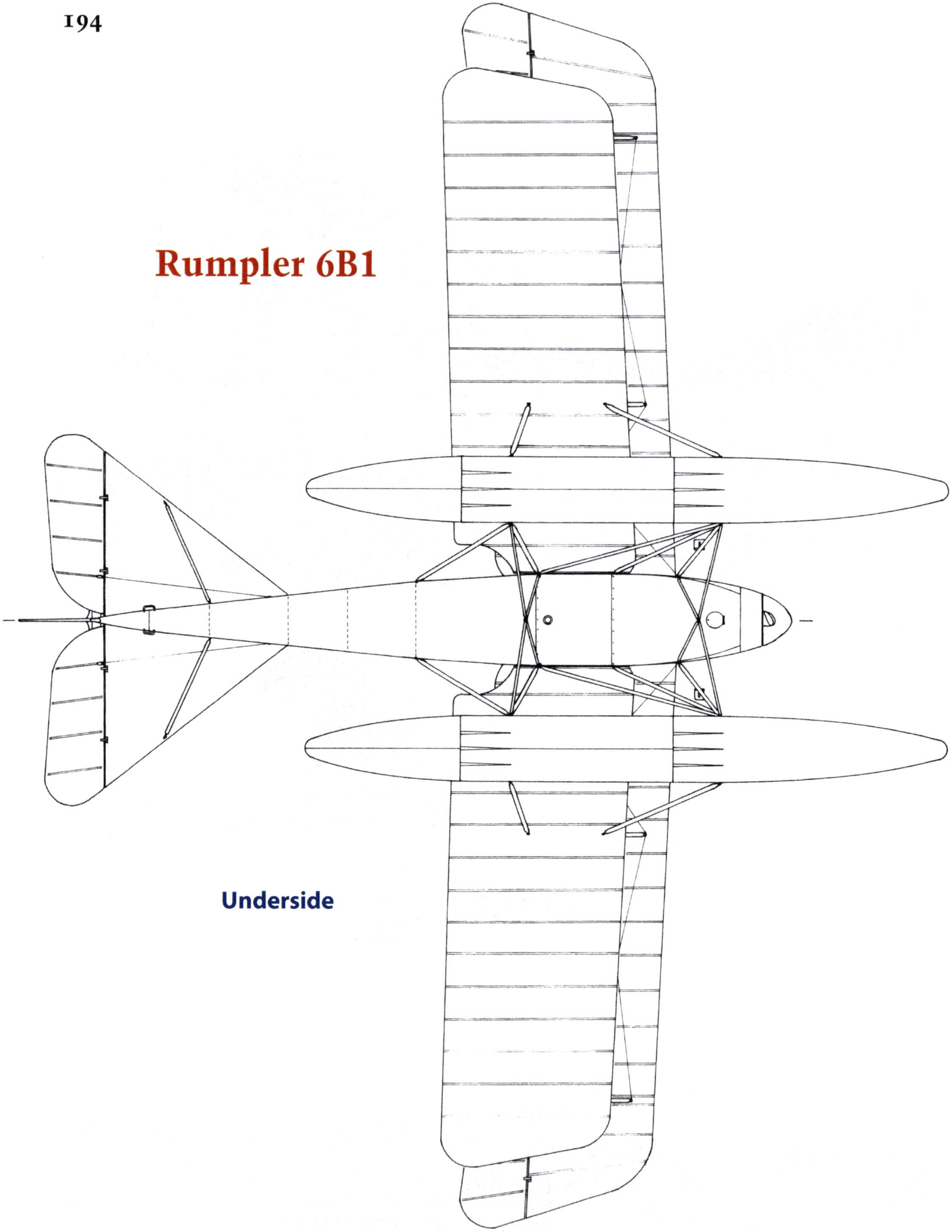

Underside

Rumpler 6B1

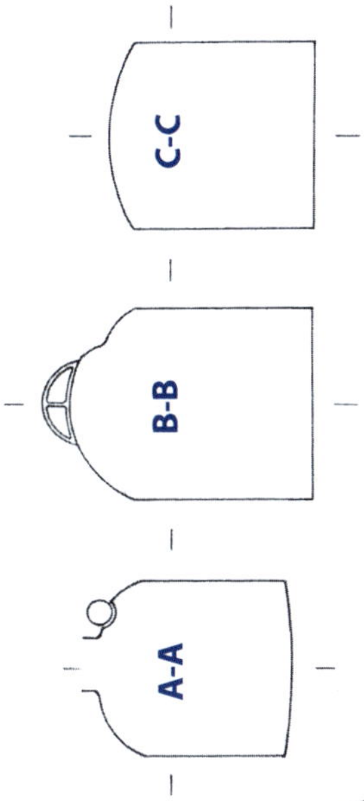

Top view with upper wing removed to show details

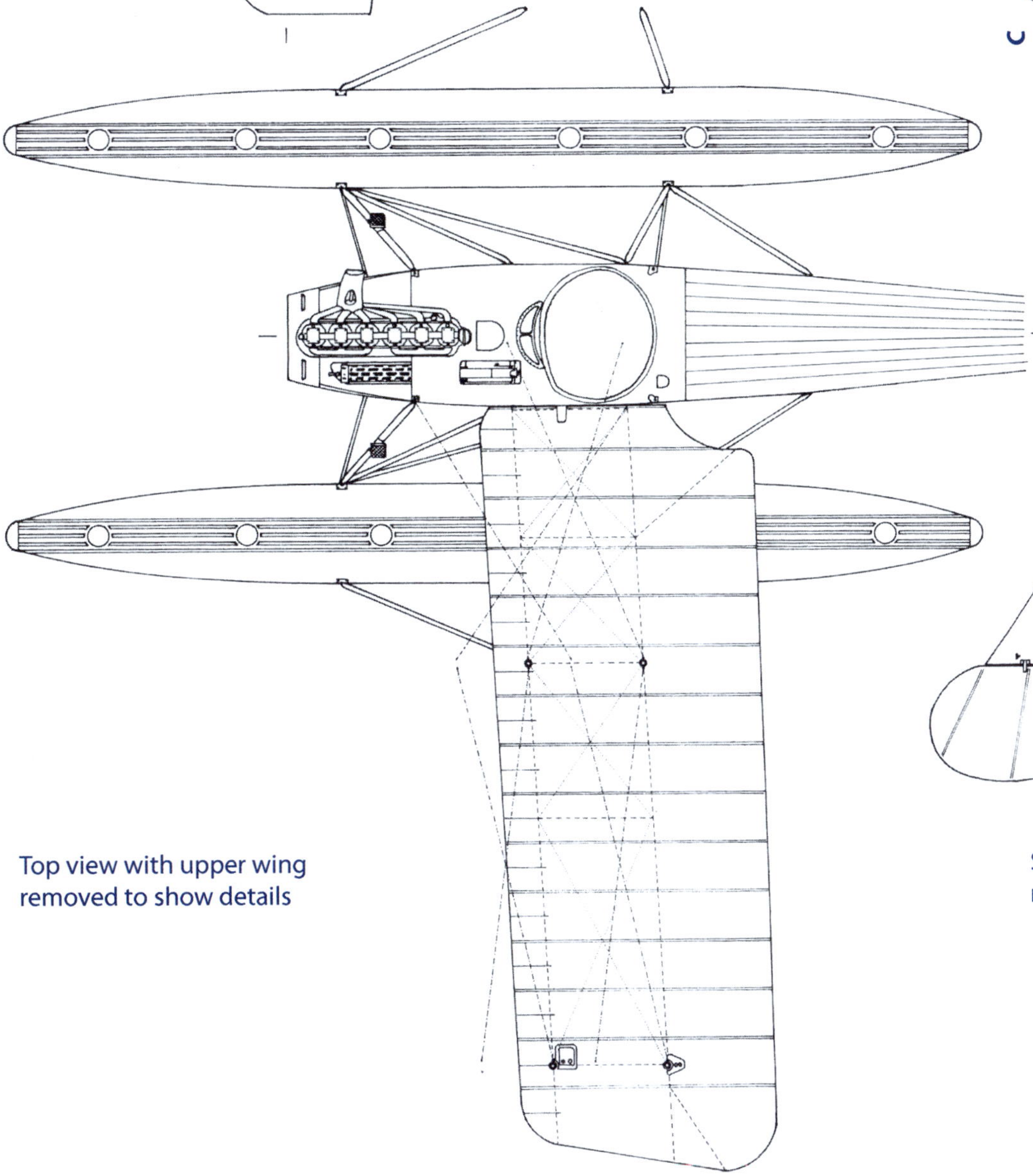

Top view with upper wing removed to show details

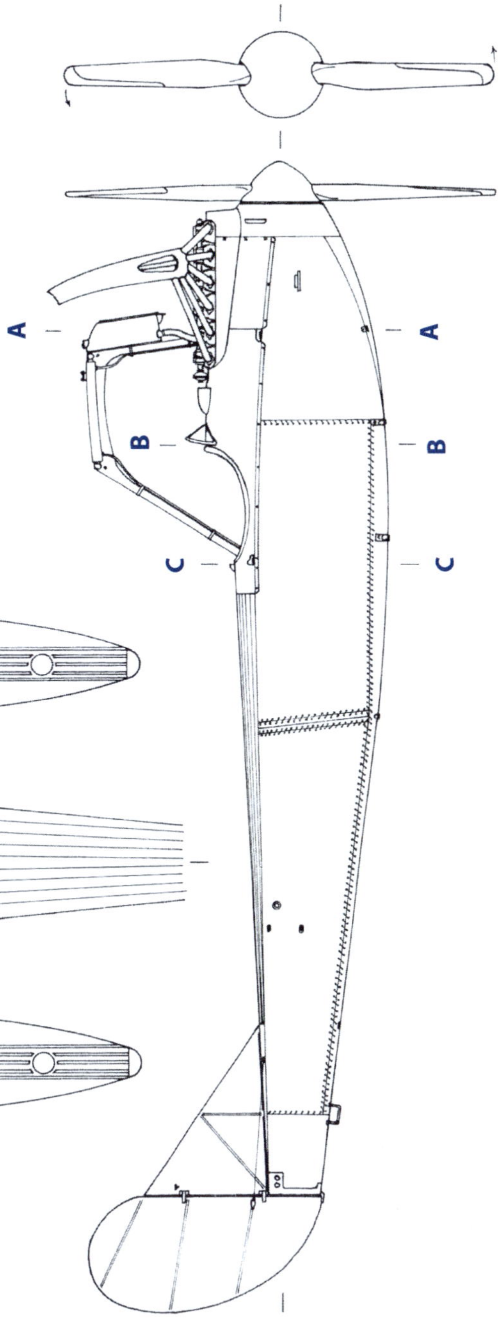

Side view with wings removed to show details

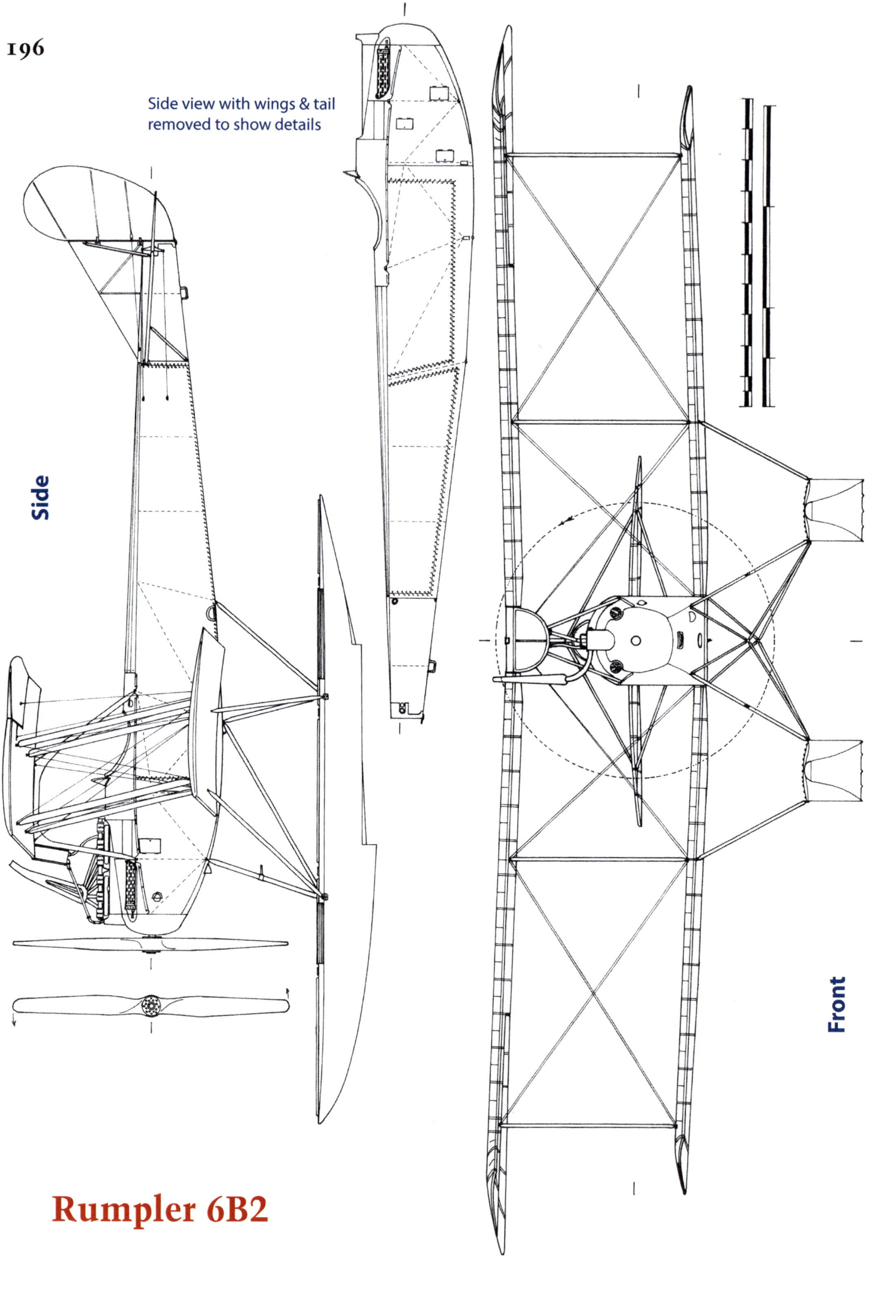
Side view with wings & tail removed to show details
Side
Front

Rumpler 6B2

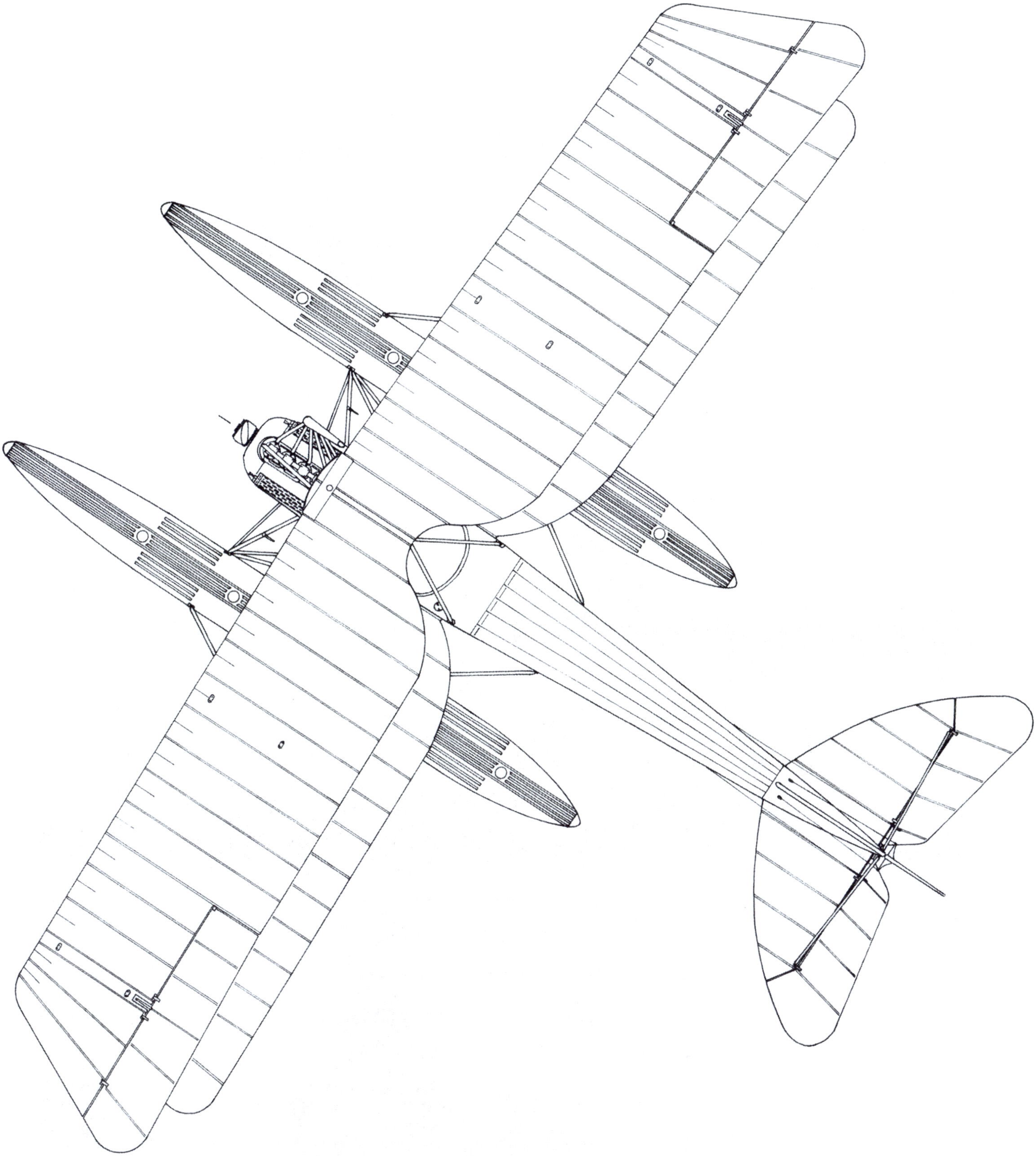

Plan View

Rumpler 6B2

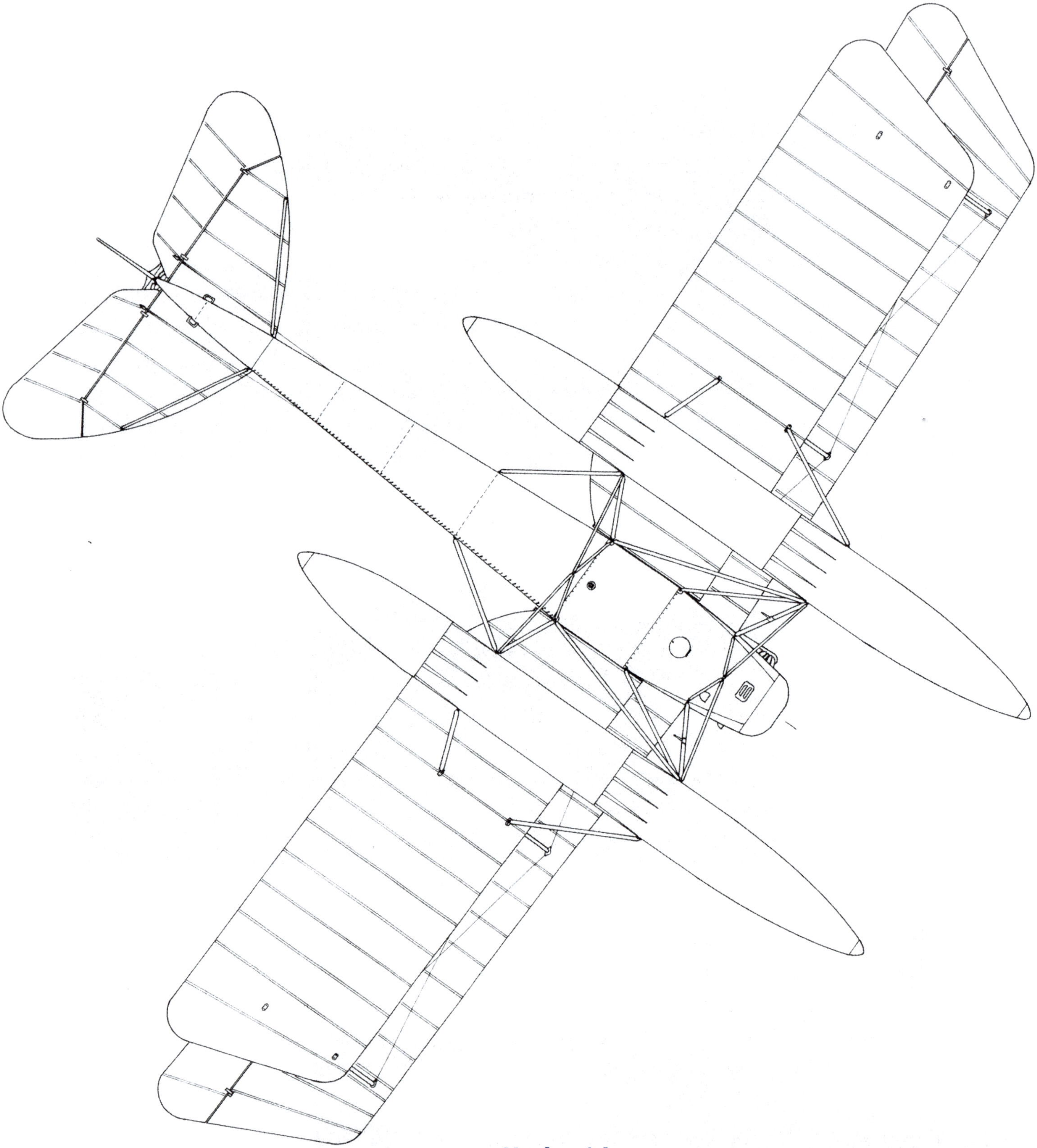

Underside

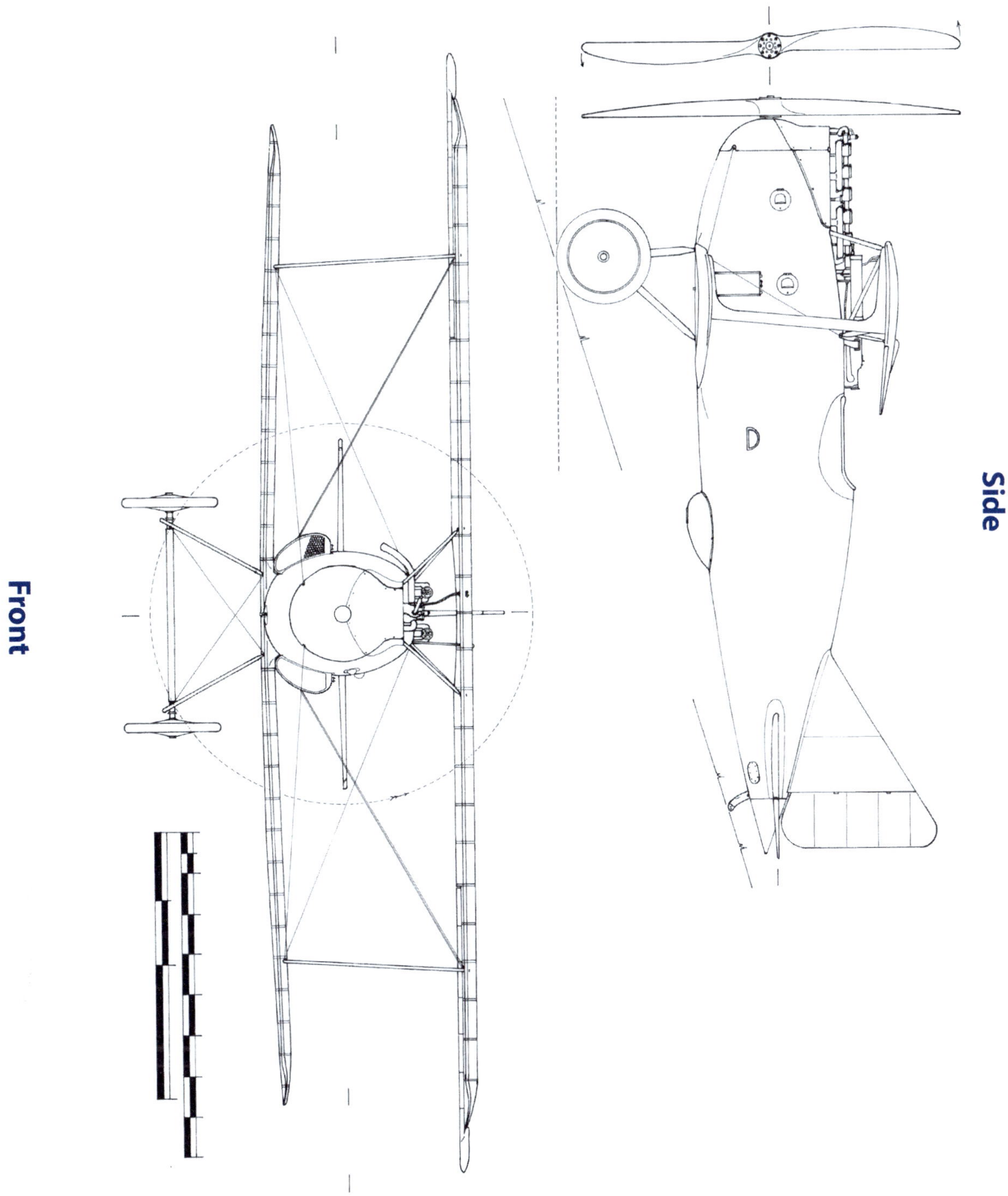
Front
Side

Rumpler D.I

D.I details with upper wing removed

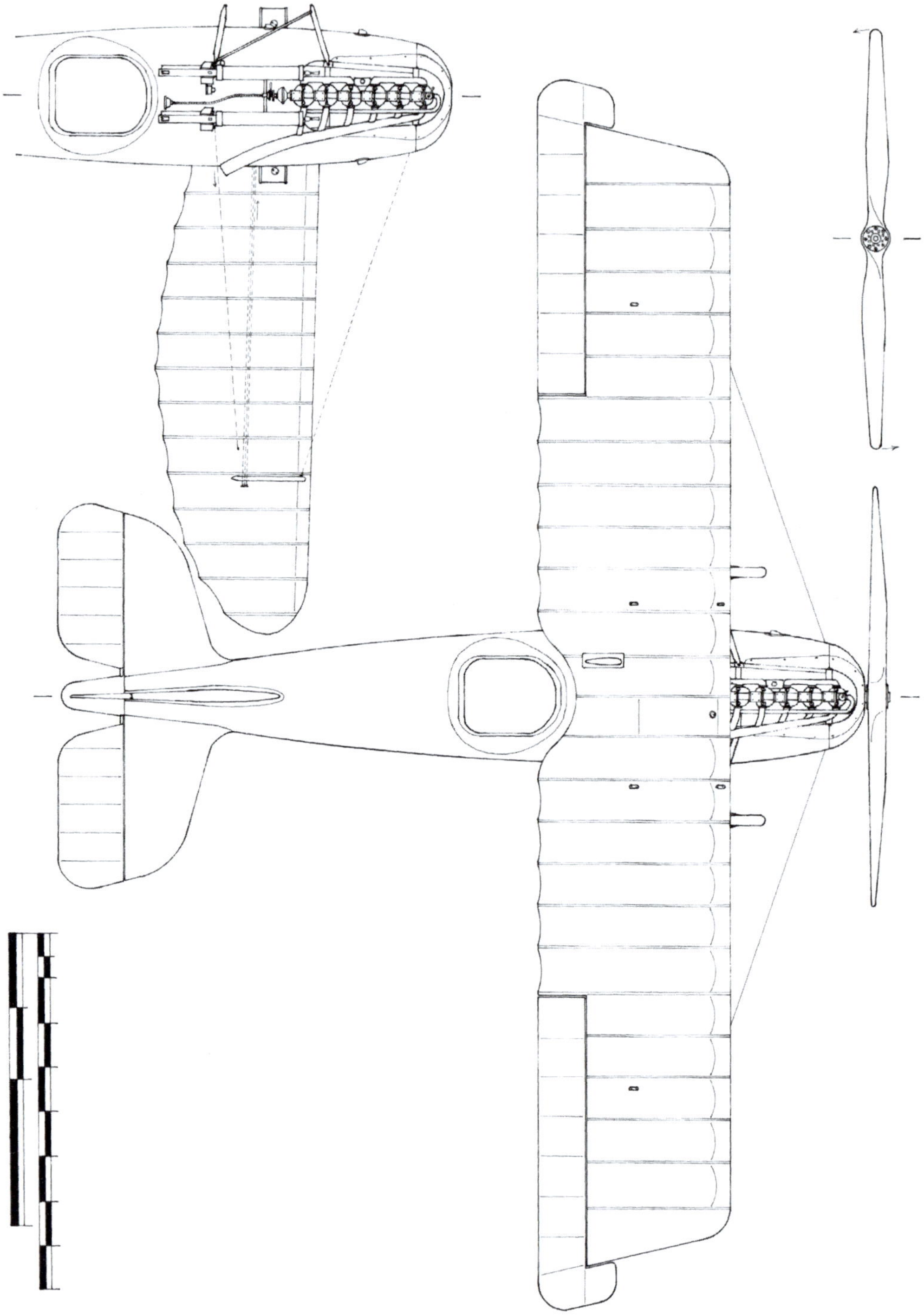

Plan View

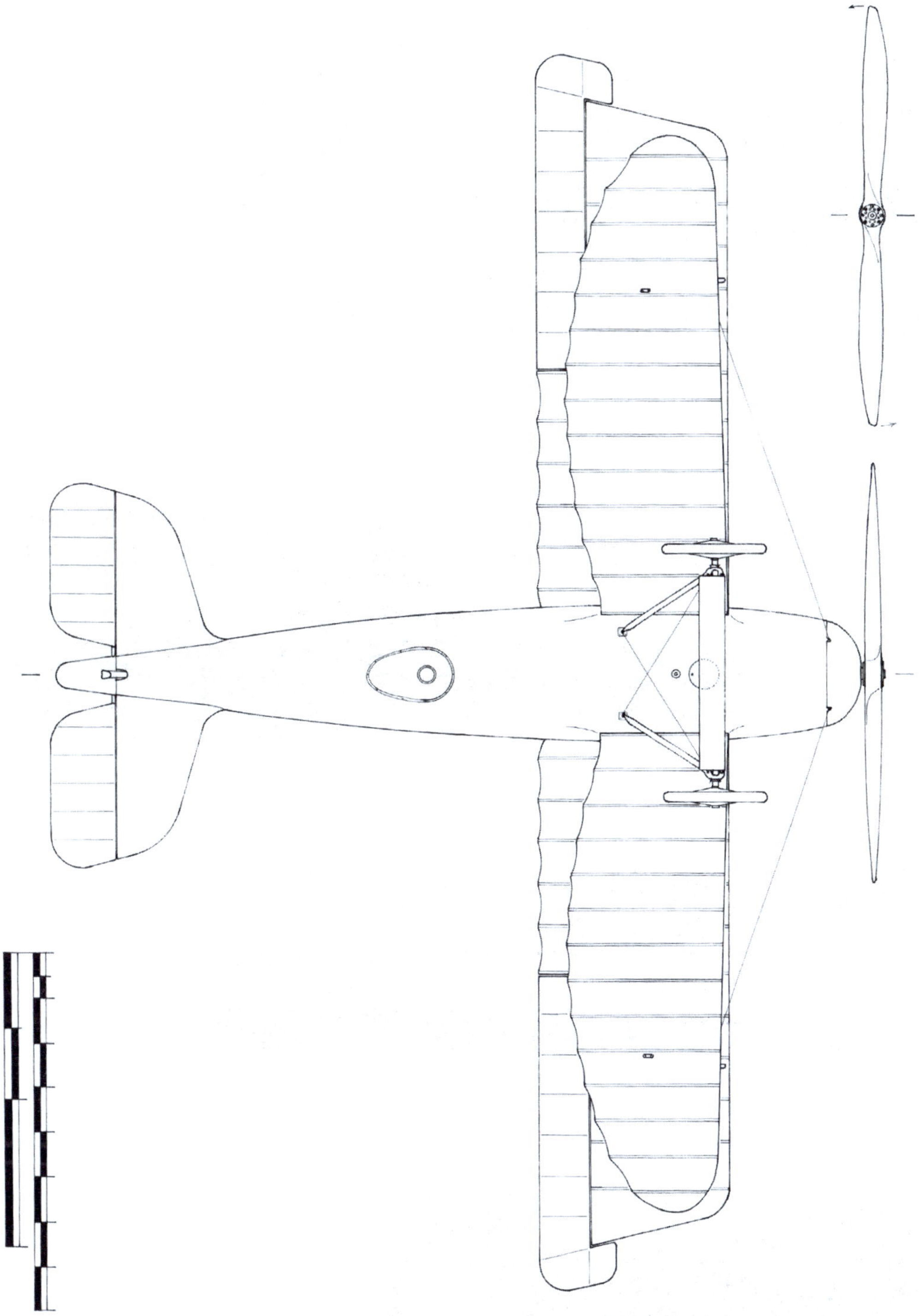

Underside

Rumpler G.I & G.II

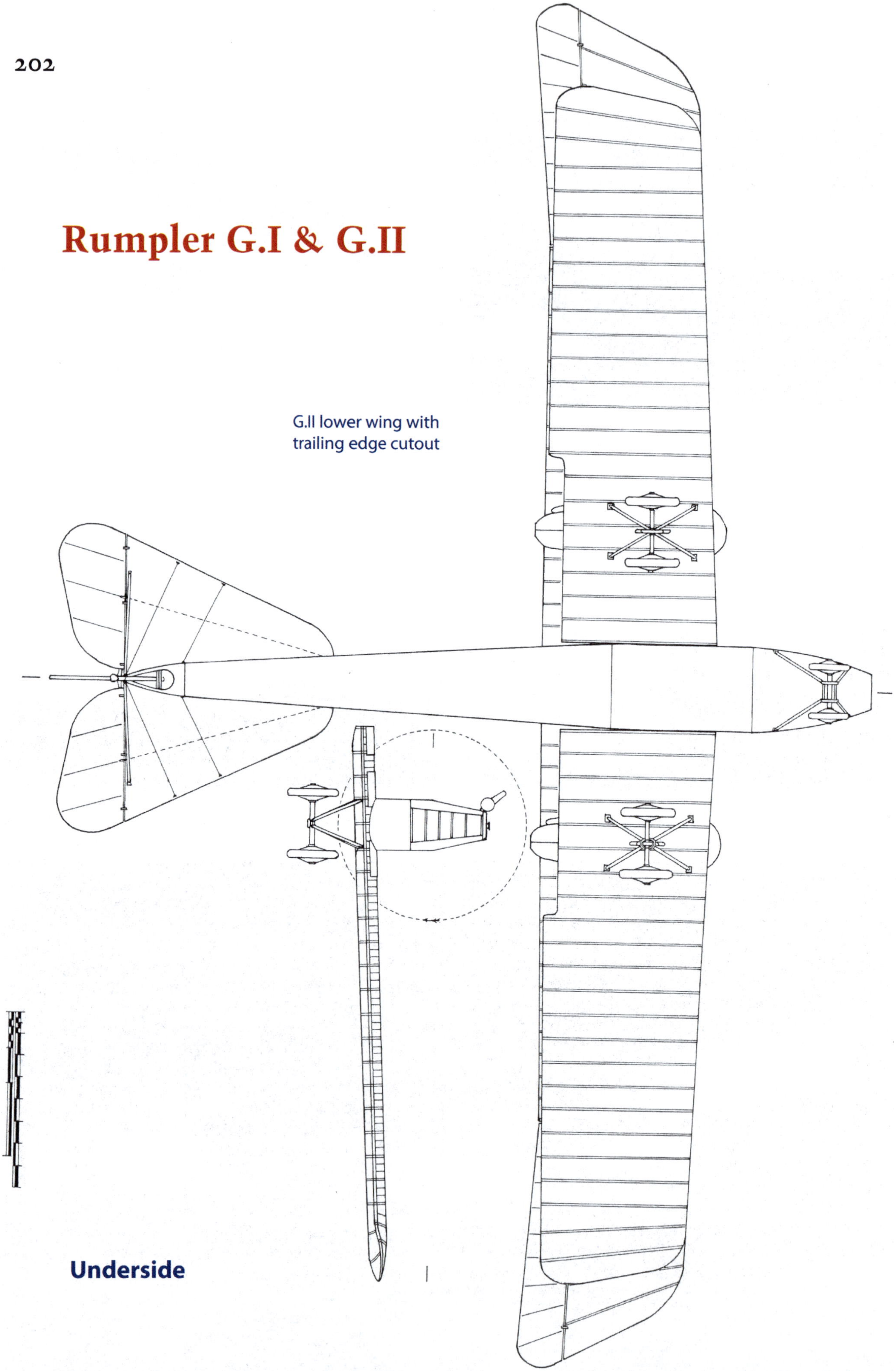

Underside

Rumpler G.I & G.II

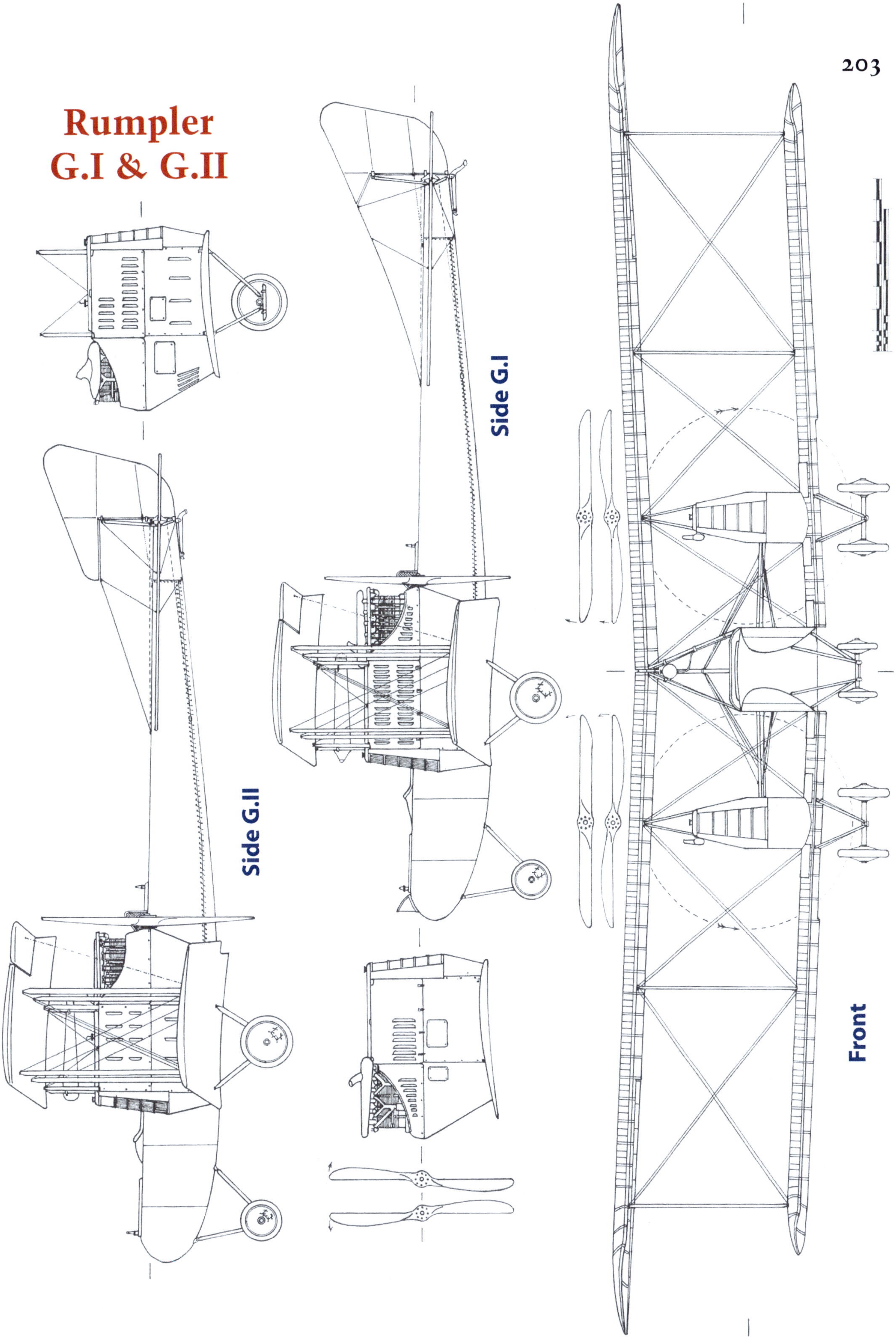

Rumpler G.I & G.II

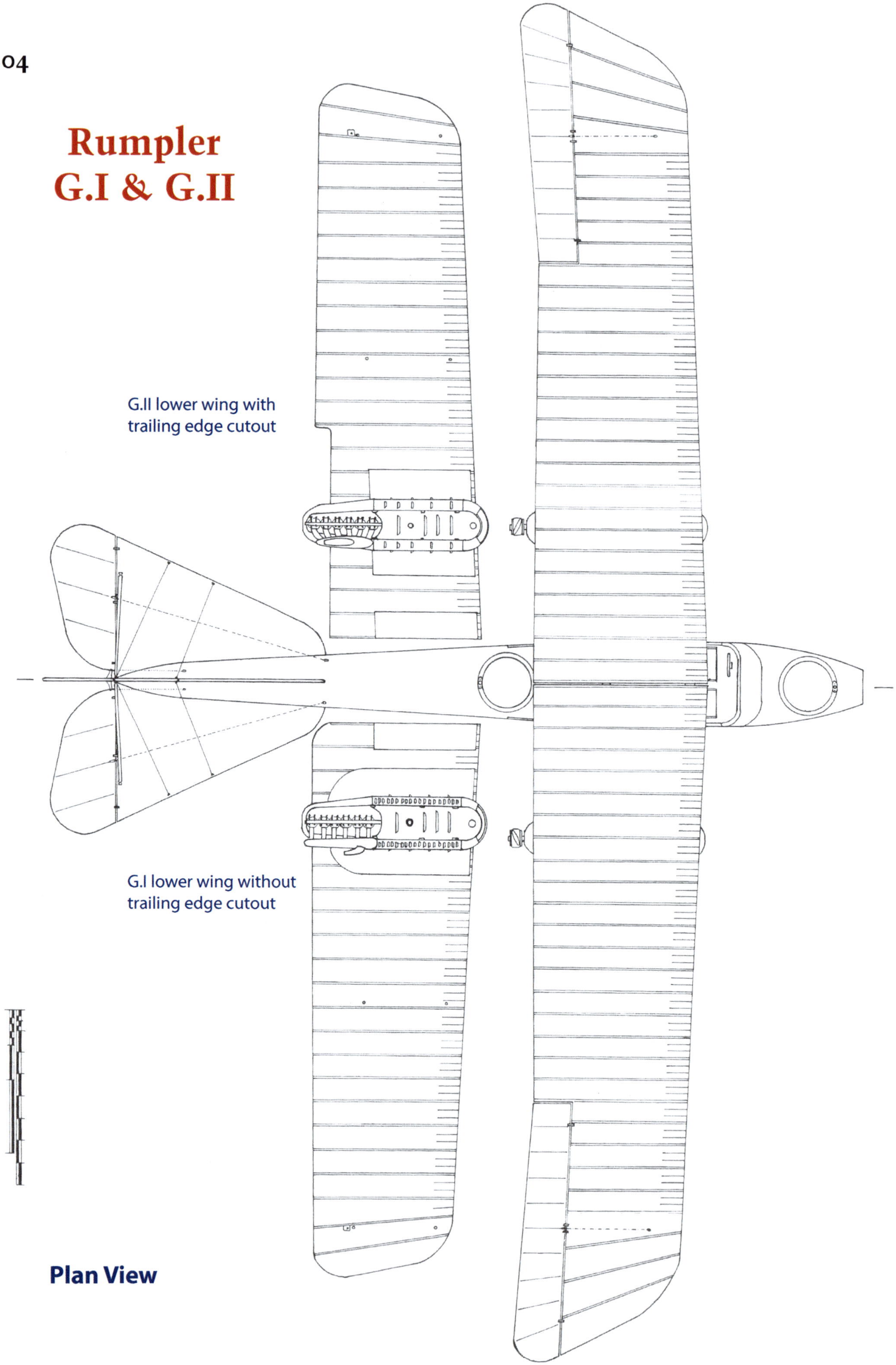

Plan View

Rumpler G.III

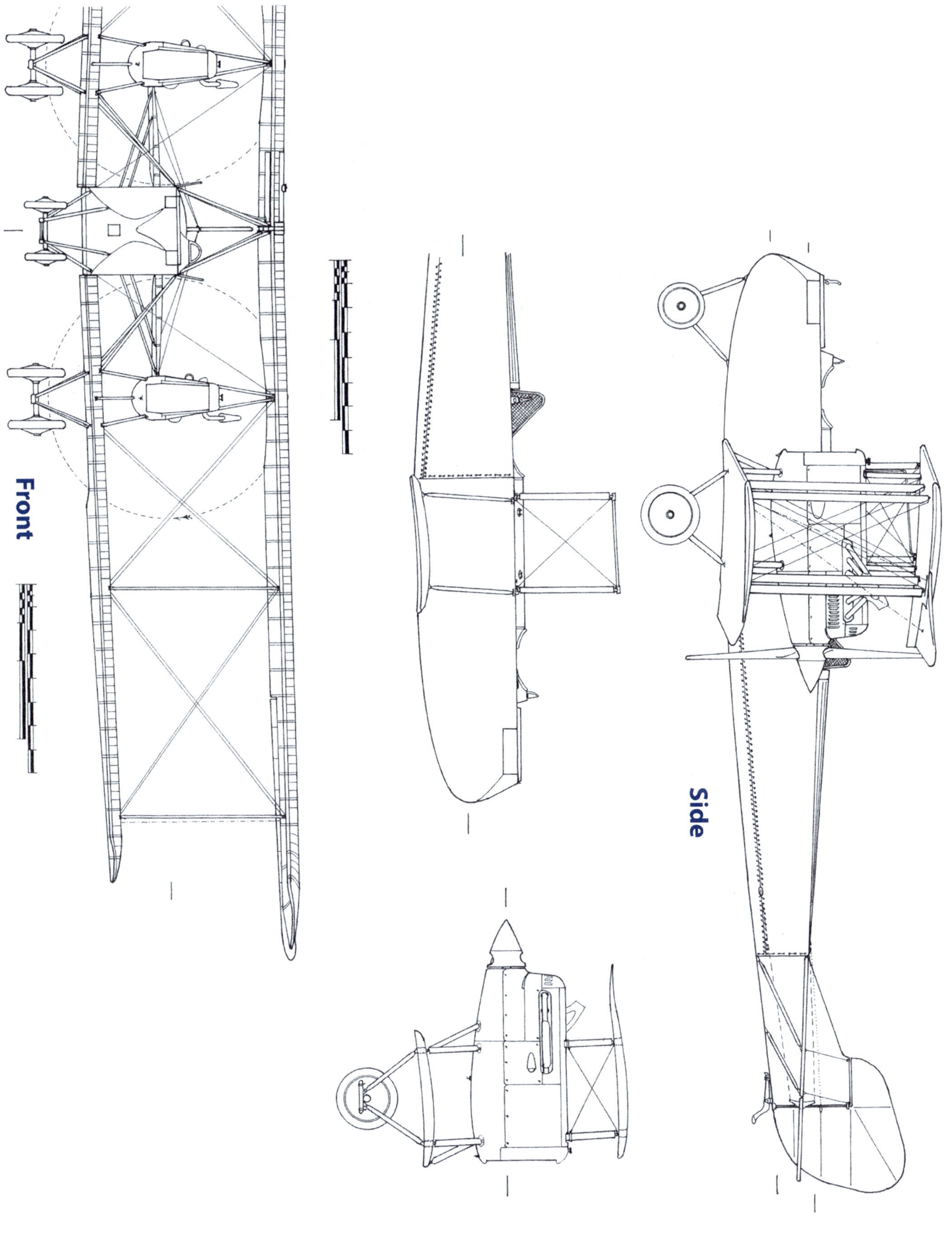

Plan View

Rumpler G.III

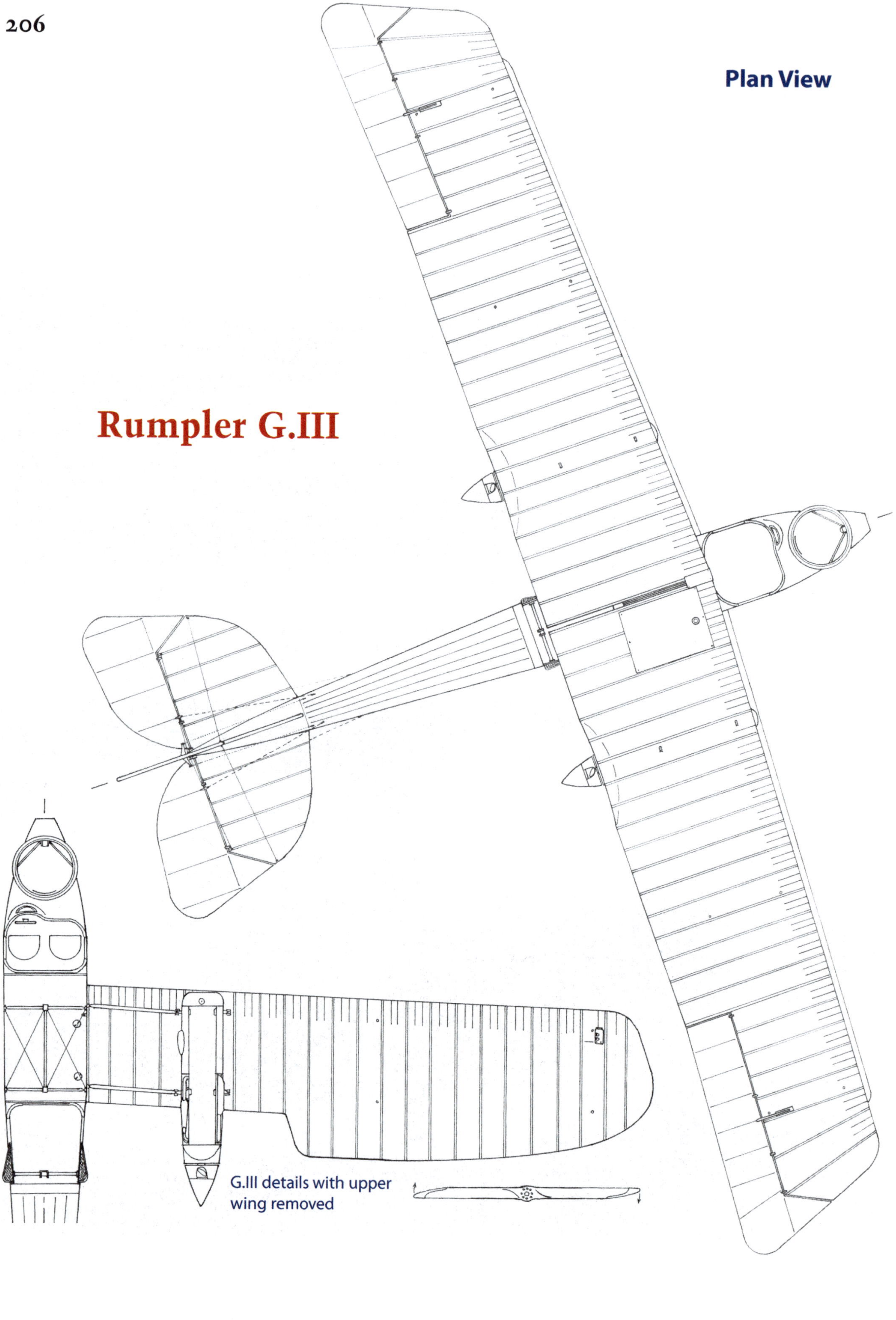

G.III details with upper wing removed

Underside

Rumpler G.III

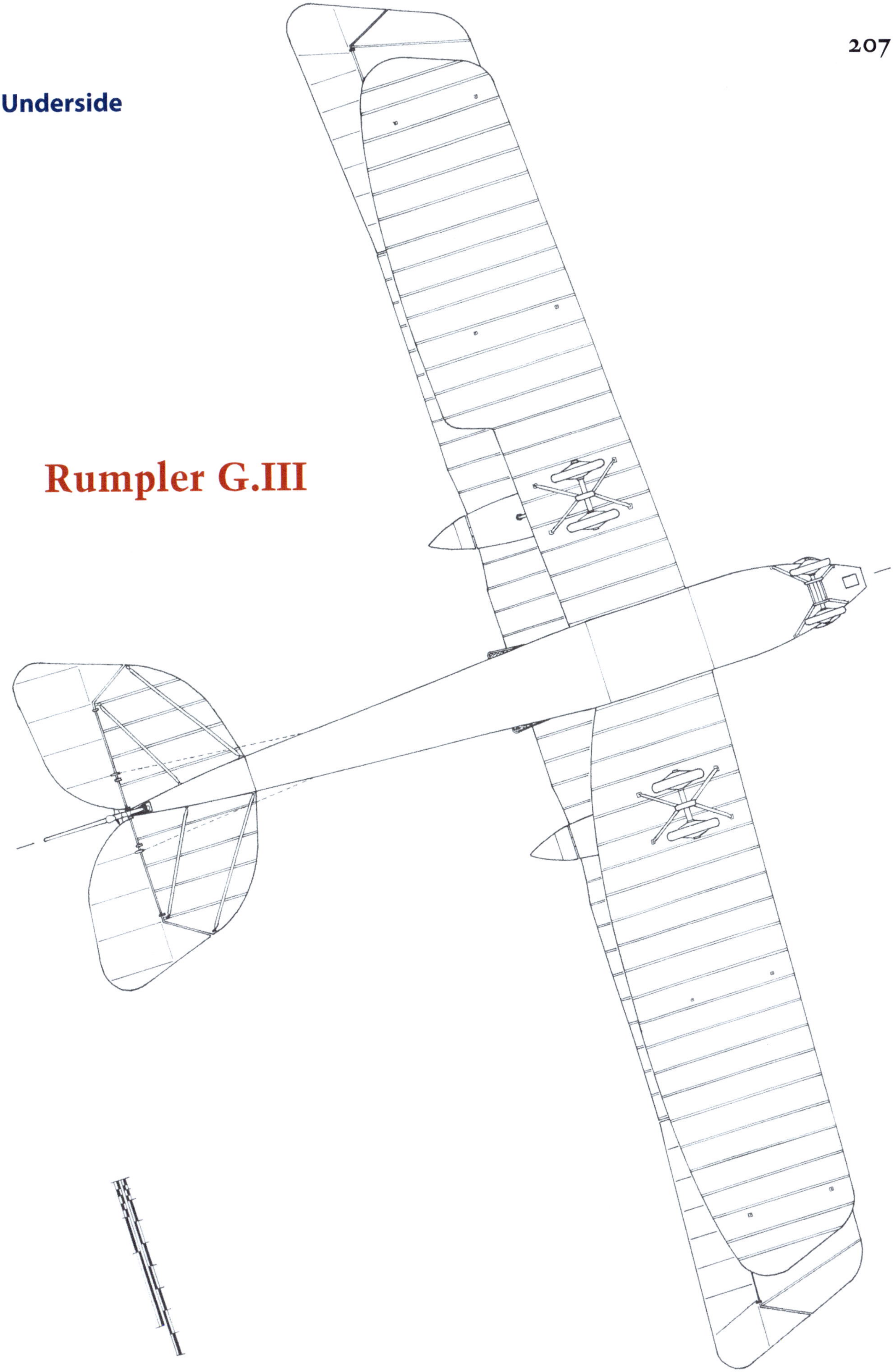

Afterword

On 25 September 1919, Ernst Udet was granted his civil pilot's license No 172, which said he was employed by the Bayerische Rumpler Werke at Augsburg. Otto Meyer, the active director, desired to have a "Rumpler Flying Day" at Augsburg to raise funds for German POW's and for publicity. On a fine warm Autumn Sunday, Udet sat in his all-red Rumpler D.I awaiting the takeoff signal. He took off and wowed the crowd with his usual low-level aerobatics. Then Robert *Ritter* von Greim took off and gave his flight demonstation. Then Udet and von Greim fought a mock combat, landing at 4:30PM.

After this a parachutist named Tony Ficklscherer made a jump from a Rumpler C.I and joyrides were given. As the crowd was dispersing , Ficklscherer (a 22 year-old pilot himself) took off in Udet's all red D.I to show what he could do. Tragically, on his second loop he hit a power magazine, the fuel tank exploded, and he was killed.

On Sunday 12 October, another display was given by Udet and von Greim at Fürth airfield attended by 15,000 spectators.

Above: *Oblt.* Robert *Ritter* von Greim is shown here before he was awarded the *Pour le Mérite*. Born in Bavaria, he scored 28 confirmed victories and commanded *Jasta* 34b. He survived the war to rejoin the new *Luftwaffe* in 1934 and was acting leader of the *Luftwaffe* in the war's last days.

Left: Robert *Ritter* von Greim is shown here in his personal Rumpler D.I, civil registration D-288, in which he gave flight demonstrations postwar. The fuselage was silver with two red stripes, the colors he used on his fighters during the war, and the word "Rumpler" was apparently in black.

Printed in Dunstable, United Kingdom

91342633R00118